Lesser known crop plants of the South Pacific:

an annotated bibliography

AF855653

by

John Palmer

Crop Research Division
Department of Scientific and Industrial Research
Lincoln

CRD

CROP RESEARCH DIVISION REPORT Nº 133
LINCOLN - 1989
ISSN 0110 -7725

Published by Crop Research Division, DSIR, Private Bag, Christchurch, New Zealand.
Printed by Caxton Press, Christchurch, New Zealand.

CATALOGUING IN PUBLICATION

PALMER, John

Lesser known crop plants of the South Pacific : an annotated bibliography / by John Palmer. - Christchurch : DSIR, Crop Research Division, 1989.

(Report / Crop Research Division, ISSN 0110-7725 ; 133).

ISBN 0-477-02551-X.

I. Title. II. New Zealand. Crop Research Division. III. Series.

UDC 016:633/635(9)

Introduction

The object of this compilation is to record some basic information sources on crop species present, but currently little utilised, in the area extending north-west from New Zealand to Papua New Guinea and east to Tahiti. There is a natural bias in information towards Fiji and Southeast Asia due to the author's background. Allowance should be made for this. Further only crops used for food, medicine or textiles are included. Forage and timber species are mentioned only incidently.

Plant introduction to the South Pacific has been a continuous process since pre-historic times with major inputs from early Melanesian and Polynesian migrants. However, all previous plant introductions were dwarfed both in number and economic importance by European imports of the past two centuries. Many species introduced with the hope of commercial exploitation have failed in this respect but remain established in the area and could be re-examined as a commercial proposition. Traditional crop species are in many instances rare and some could need protection. These plants have been emphasised where only minimal references to more common species are included.

For their assitance in the preparation of this work I have to thank, in particular, Bill Sykes who read the script and adjusted my nomenclature, Robert Lamberts for taking the better photos; the less clear ones are my own, and Bruce Muschamp for his expertise in preparing the script for the laser printer. The drawings are copies of some of an unpublished set of Indian crop plants collected some fifty years ago from the India Museum, London.

In addition to the quoted references the following publications have been perused for vernacular or common names:

Jardin, C. (1974). Kulu, kuru, uru; lexicon of names of food plants in the South Pacific. South Pacific Commission, Noumea. 203pp.

Ohwi, J. (1965). Flora of Japan. Smithsonian, Washington. 1967pp.

Parham, B.E.V. (1942). Fijian plant names. Department of Agriculture, Suva. 83pp.

Parham, B.E.V. (1972). Plants of Samoa. A guide to their local and scientific names. DSIR Information Series no. 85, Wellington. 162pp.

Steiner, M.L. (1960). A dictionary of vernacular names of Pacific food plants. National Research Council of the Philippines. 371pp.

John Palmer

John Palmer, Lincoln. January 1989

Figure 1: Okra plant.

Abelmoschus esculentus (L.) Moench

Malvaceae

Common
Okra , Ladies finger, Gumbo, Bhindi (Hindi), Ura ura (New Guinea), Aikavitu (Fiji).

Synonyms
Hibiscus esculentus L.

Notes
Probably a polytypic complex and not a single species. Thought to have originated in Africa but cultivated in India for several millennia. Now pan tropical. The mucilaginous seed pod is particularly liked by people of Indian extraction. Okra is strictly a tropical vegetable and can only be grown outdoors in northern N.Z. with difficulty.

References
Palmer, J. (1972). Okra. Crop Research News (DSIR) No. 9, 17–18.

Martin, F.W. and Ruberte, Ruth (1978). Vegetables for the hot humid tropics, Pt 2, Okra. U.S. Dept. Agric. New Orleans, 22 p. (33 references).

Abelmoschus manihot (L.) Medic.

Malvaceae

Common
Sunset hibiscus (U.S.A.), Aibika (New Guinea), Mbele (Fiji), Pele (Polynesia).

Synonyms
Hibiscus manihot L.

Notes
A coarse perennial, suffruticose shrub, 1 to 3 m high, widely cultivated in Malesia and introduced to Polynesia at an early date. An important leaf vegetable in Papua New Guinea and many islands of the South Pacific. There are at least 4 recognised cultivars in Fiji and a wide range of leaf shapes in Indonesian plants.

References
Westwood, V. and Kesavan, V. (1982). Traditional leafy vegetables of Papua New Guinea : Aibika. Proceedings of the Second Papua New Guinea Food Crop Conference, Port Moresby. pp 464–473.

Ochse, J.J. (1977). Vegetables of the Dutch East Indies. Australian National University Press, Canberra. pp 464–473.

Abelmoschus moschatus (L.) Medic.

Malvaceae

Common
Musk mallow, Ambrette, Latakasturika (Sanskrit), Fautia (Tahiti), Waki waki (Fiji), Aute toga (Samoa).

Synonyms
Hibiscus abelmoschus L.

Notes
A shrubby herbaceous perennial (to 3 m) from India, southern China, through Malesia to the Pacific. Doubtfully native in Fiji but widespread and naturalised. Cultivated in Java, Martinique and the West Indies for the aromatic seeds from which musk oil or ambrette is extracted for perfume.

References
The Wealth of India – Raw Materials Vol. 4. CSIR, New Delhi. pp75–77 (under Hibiscus)..

Poucher, W.A. (1979). Perfumes, Cosmetics and Soaps. 7th edition. Chapman and Hall, London. p29.

Acer saccharum Marshall

Aceraceae

Common
Sugar maple.

Synonyms
A. saccharinum Wangenh. non L.

Notes
The main species tapped for maple sugar in eastern North America. Trials have recently begun in N.Z.

References

Olsen, J.B. (1983). Sugar maple development. N.Z. Tree Grower 4(4), 86–87.

Creasy, R. (1986). The gardener's handbook of edible plants. Sierra Club Books, San Francisco. pp 163–166.

Greene, S. (1954). Sugar weather in the green mountains. National Geographic Magazine 105, 471–482.

Aegle marmelos (L.) Correa

Rutaceae

Common
Bengal quince, Golden apple, Bael, Bel (Hindi).

Synonyms
Crataeva marmelos L.

Notes
A sub-tropical fruit tree indigenous to dry hill areas of India and Southeast Asia. Cultivated and occasionally naturalised on some larger islands of the South Pacific. Will not fruit without a dry season. The fruit is globular, some 6 cm. in diameter with a hard yellow to orange skin. The sweet insipid flesh is used in drinks and sherbets. Bael also has many uses in local medicine.

References

Morton, J.F. (1987). Fruits of warm climates. Creative Resource Systems, Winterville. pp 187–190.

Allen, B.M. (1967). Malayan fruits. Donald Moore Press, Singapore. pp 169–172.

Roy, R.S. and Singh, R.N. (1979). Bael fruit *(Aegle marmelos)* – a potential fruit for processing. Economic Botany 33 (2), 203–212.

Aleurites fordii Hemsley

Euphorbiaceae

Common
Tung oil, Chinese wood oil, Tung–yu (China).

Notes
A tree native to the warm temperate areas of China. Seeds exported mainly from Hankow for oil extraction. This is used in paint and varnishes. Tung was widely cultivated between the two world wars including in N.Z. and Florida, but is not now known to be grown commercially outside China. N.B. Some tung trees are still growing at Te Karaka near Gisborne, at Napier and in Northland. Vide: Pollock R.P. and Halliwell, G. (1978). A guide to some lesser known tree and other crops, mainly for the warmer parts of N.Z. MAF, Gisborne, p. 2.

References

"G.F." (1933). The cultivation of the tung tree in N.Z. J. N.Z. Inst. Hort. 3(4), 87–91.

Burns, M.M. et al. (1939). A survey of tung groves in N.Z. DSIR Bull. 66, Wellington. 61 pp.

Li Lai–yung (1943). The influence of stratification of tung seeds upon emergence and establishment of seedlings in a nursery. N.Z. J. Sci. Technol 25A(1), 43–48.

Aleurites moluccana (L.) Willd.

Euphorbiaceae

Common
Candle nut, Kukui (Hawaii), Tiairi (Tahiti), Lama (Samoa), Lauci (Fiji), Tuitui (Polynesia).

Synonyms
Aleurites triloba Forst.

Notes
A tall monoecious tree bearing spherical fruit some 3 cm in diameter. A hard green skin covers the thick wrinkled shell of the nut. The endosperm is sufficiently oily to burn readily and was at one time an important source of illumination in Polynesia. The nut was also eaten roasted but could be toxic in large quantities. Oil was extracted from the nuts on a commercial scale during the 19th century in Tahiti and in Hawaii and an analysis is given by Petard (1986).

References

Degener, O. (1945). Plants of Hawaii National Park illustrative of plants and customs of the South Seas. Edwards Brothers, Ann Arbor. pp 193 – 199.

Petard, P. (1986). Plantes utiles de Polynesie. Edition's Haere Po No Tahiti. pp 195 – 198.

See also: Safford, W.E. (1905). The useful plants of the island of Guam. Contributions from the United States Herbarium, Vol. 9, p 177.

Allium chinense G. Don

Liliaceae

Common
Rakkyo (Japan), Ch'iao t'ou (China).

Synonyms
A. bakeri Regel, *A. splendens* sensu Miq.

Notes
A little known bunching onion from the Far East, bulbs commonly pickled like shallots but of crisper texture. Plants go dormant in summer. Flowering is said to be sparse and seed sterile.
Other Chinese/Japanese Allium spp. occasionally cultivated include: *A. grayi* Regel (syn. *A. nipponicum* Fr. & Sav.) No–biru (Japan) young leaves and bulbs eaten in spring. *A. victorialis var. platyphyllum* (Hult.) Makino, Gyoja–ninniku (Japan). *A. monanthum* Maxim. Hime–nira (Japan). N.B. *Allium ledebourianum* Roemer & Schultes f. is a synonym of *A. schoenoprasum var. foliosum* Regel, an Eastern variety of the chive.

References

Mann, L.K. and Stearn, W.T. (1960). Rakkyo or ch'iao t'ou (*Allium chinense* G. Don syn. *A. bakeri* Regel). Economic Botany 14 (1), 69–83.

Kato, N. (1968). Herbs used in northern Japan. Plants and Gardens 24(2), 18–19.

Yashiroda,K. (1968). Dictionary of popular Japanese herbs. Plants and Gardens 24 (2), 44–58.

Allium fistulosum L.

Liliaceae

Common
Japanese bunching onion, Welsh onion, Negi (Japan).

Notes
A non-bulbing onion, cultivars differ in their tendency to form vegetative offsets. Commonly cultivated in N.Z. as 'spring onions'. The main garden onion of China since prehistoric times.

References

Stearn, W.T. (1943). The welsh onion and ever-ready onion. Gardeners' Chronicle Ser 3, 2958, 86–88.

Herklots, G.A.C. (1972). Vegetables in Southeast Asia. Allen & Unwin, London. pp 396–397.

Allium schoenoprasum L.

Liliaceae

Common
Chives

Synonyms

Notes
A widespread species in Europe and Northern Asia. The cultivation of chives is important in Europe, particularly Germany, where winter forcing under cover is common. In N.Z. small quantities are grown in summer for processing.

References

Folster, E. and Krug, H. (1977). Influence of the environment on growth and development of chives. 2. Breaking the rest period and forcing. Scientia Horticulturae 7(3), 213–224.

Thompson, Wendy A. and Lammerink, J. (1983). Chive cutting for processing. N.Z. Commercial Grower 38(4), 23.

Allium tuberosum Rottler

Liliaceae

Common
Chinese chives, Chinese garlic, Nira (Japan), Gow Choy (China)

Synonyms
Allium odorum L., *A. uliginosum* Don., *A. tataricum Ait.*

Notes
A small non bulbing onion commonly used as flavouring in the Far East. The lower stems are blanched individually like small leeks and immature flower heads are also marketed.

References
Yakawa T. (1961). Studies on tillering in the genus Allium 6. On tillering of Chinese leek (*Allium tuberosum* Rottler) under natural conditions. J. Jap. Soc. Hort. Sci. 30, 299–310 (English summary).

Herklots, G.A.C. (1972). Vegetables in Southeast Asia. Allen and Unwin, London. Pp. 399–401.

Alocasia macrorrhiza (L.) Schott

Araceae

Common
Giant Taro, Ape (Tahiti), Via gaga (Fiji), Biga (Philippines), Kape (Cook Is), Tahmu or Ta'ama (Samoa)

Synonyms
Alocasia indica Schott (?); *A. indica* C. Koch; *Colocasia macrorrhiza* Schott.

Notes
A traditional, but minor, aroid root crop of the Pacific Islands. Widespread but now largely uncultivated and a weed on some sub–tropical islands. A large plant 1.8 metres or more in height. The thick epigeous stem is the edible portion.

References
Furtado C.X. (1984). *Alocasia macrorrhiza* and its varieties. Gardens Bulletin (Singapore) 11(3), 244–257.

O'Hair, S.K. and Asokan, M.P. (1986). Edible aroids: botany and horticulture. Horticultural Reviews 8, 43–99. Avi Publ. Westport, Conn.

Aloe vera (L.) Tourn.

Liliaceae

Common
Aloe, Bitter aloes

Synonyms
Aloe barbadensis Mill.

Notes
The major medicinal aloe, cultivated in the West Indies, Florida, Texas and Queensland. Other species utilised include *A. ferox* Mill. and *A. africana* Mill. from South Africa and *A. parryi* Baker from the Arabian Gulf area.

References
Morton, Julia F (1961). Folk uses and commercial exploitation of aloe leaf pulp. Econ. Bot. 15, 311–319.

Grindlay, D (1985). *Aloe vera*. The Garden 110 (11), 534–535.

See also: Letter to the editor by Forster P.I. (1985). Queensland Agric. J. 111 (5), 232 suggesting that two varieties of *A. vera* occur with either yellow or orange flowers.

Althaea officinalis L.

Malvaceae

Common
Marshmallow

Notes
A herbaceous perennial native to saline areas of northern Europe. Cultivated in Belgium, France and Germany. Rare in New Zealand. Two year old roots contain 15 to 20% mucilage which is used in medicine and to a limited extent in the food industry. A small scale trial at Lincoln 1980 – 83 yielded less than the equivalent of 1000 kg dried root per hectare and suffered from a root rot in the third winter.

References
Morton, Julia F. (1977). Major medicinal plants. Charles C. Thomas, Springfield. pp 206 – 208.

British Herbal Pharmacopoeia (1983). British Herbal Medicine Association, Keighley pp 23 – 24.

See also: Ilieva St, and Peneva ,P. (1984). A new variety of *Althaea officinalis* L. Acta Horticulturae No. 144, 67 – 70.

Figure 2: Marshmallow.

Figure 3: Amaranthus hypochondriacus.

Amaranthus spp.

Amaranthaceae

Common
Grain amaranths, Kiwicha (Quechua name for *A. caudatus*).

Notes
Three species grown for grain production are:– *A. caudatus* L. from the Andes, Southern Ecuador to Peru. *A. hypochondriacus* L. from Mexico. *A. cruentus* (L.) Sauer from South Mexico & Guatemala. Recent interest has been sparked in the grain amaranths as the seed is relatively high in protein, especialy the lysine fraction, as compared with the major cereal grains. Numerous trials have been conducted in the U.S.A. and seed of most, if not all three species, has been imported into N.Z. since 1980 (vide R. Crowder, Lincoln College). Mechanical harvesting is hampered by late flowering and uneven seed maturity in N.Z. *A. hypochondriacus* is the principal grain amaranth of Northern India where it has been grown for centuries.

References
Vietmeyer, N.D. (1984). Amaranth. Modern prospects for an ancient crop. National Academy Press, Washington. 80 pp.

Kauffman, C.S. and Haas, P.W. (1982). Grain amaranth: an overview of research and production methods. Rodale Research Report NC–83–5, Rodale Press, Emmaus. 13 pp.

See also: Marx, Jean L. (1977). Amaranth, a comeback for the food of the Aztecs? Science 198 (4312), 40.

Amaranthus tricolor L.

Amaranthaceae

Common
Amaranth, Tampala spinach (U.S.A.), Tumbua (Fiji), Chauria (Fiji Indian), Yin Choi (Hawaii). Ornamental, coloured leaved cultivars are sometimes called 'Joseph's coat.'

Synonyms
Amaranthus gangeticus L., *A. melancholicus* F. Vill., *A. oleraceus* Burm. f., *A. mangostanus* Blanco.

Notes
A common leaf vegetable in South Asia from India to China. More commonly grown in Melanesia than Polynesia. Other species used as a leaf vegetable include *A. lividus L.(syn. A. blitum)* particularly in Greece, and *A. cruentus* a grain producing species from Central America. *A. viridis* L. (syn. *A. gracilis* Desf.), a common adventive in most Pacific archipelagos, and is occasionally cultivated.

References
Campbell, T.A. and Abbott, J.A. (1982). Field evaluation of vegetable amaranth. Hort Science 17(3), 407–409.

Martin, F.W. and Telek L (1979). Vegetables for the hot humid tropics. Part 6. Amaranthus and Celosia. USDA, New Orleans. 21 p.

Amoracia rusticana Gaertner

Cruciferae

Common
Horseradish.

Synonyms
Amoracia lapathifolia, Gilib.; *Cochlearia amoracia* L.; *Cochlearia rusticana* Lam.

Notes
A deep rooted perennial herb. The root is used as a condiment. Both diploid and tetraploid forms occur. Trials at Crop Research Division, Lincoln in the late 1970's were inconclusive.

References
Anon (1957). Horseradish. The farm quarterly 11(4), 66–67, 84–85.

Courtier, J.W. and Rhodes, A.M. (1969). Historical notes on horseradish. Economic Botany 23, 156–163.

Thieret, J.W. (1957). Horseradish. Utilisation abstract. Economic Botany 11, 348.

Amorphophallus konjac Koch

Araceae

Common
Elephant yam, Konjaku (Japan), Mo–yu (Chinese–Hawaii).

Synonyms
Amorphophallus rivieri Dur.

Notes
Of little economic value except in Japan. The cultivated *Amorphophallus sp.* of India and some Pacific Islands is *A. paeoniifolius* (Dennst.) Nicolson. The genus is distinctive due to the large, much dissected leaf borne on a single thick stem.

References
Kurihara, H. (1979). Trends and problems with konjak cultivation in Japan. Japan Agric. Res. Quart. 13, 174–179.

Ohtsuki, T. (1968). Studies on reserve carbohydrates of four Amorphophallus species, with special reference to marram. Bot. Mag., Tokyo 81, 119–126.

Amorphophallus paeoniifolius (Dennst.) Nicholson

Araceae

Common
Elephant yam, Daiga (pronounced ndainga, Fiji), Suran (Fiji Indian), Teve (Cook Is. and Tahiti).

Synonyms
Amorphophallus campanulatus Bl. ex Decne.

Notes
The principal Amorphophallus cultivated in India, S.E. Asia and the Pacific. Other species are utilised locally. A crop of minor importance wherever it occurs. The flattened circular tubers can be very large, sometimes over 10 kg in weight.

References
Mishra S. et al. (1983). Production and profit potential of Amorphophallus affected by seed size. Indian J. Agric. Sci. 53(1), 27–29.

Mehta, Y.R. (1959). Suran or Zimikand. Vegetable production in Uttar Pradesh. Bulletin of Agricultural Information Bureau, Lucknow. 21 p.

Figure 4: *Amorphophallus sp.*

Anacardium occidentale L.

Anacardiaceae

Common
Cashew nut, Maranon (Central America).

Synonyms
Cassuvium pomiferum Lam.

Notes
A tree of the semi arid tropics. South American in origin but western India and East Africa are major commercial sources. Introduced in some Pacific Islands, particularly Tahiti, Guam and Yap.

References
Wait, A.J. and Jamieson, G.I. (1986). The cashew, its botany and cultivation. Queensland Agricultural Journal 112(5), 254–257.

Ohler, J.G. (1979). Cashew. Communication 71, Royal Tropical Institute, Amsterdam.

Angelica archangelica L.

Umbelliferae

Common
Angelica.

Synonyms
Angelica officinalis Hoffm.

Notes
A hardy monocarpic perennial from northern Europe. The subspecies *Angelica archangelica spp. archangelica* mainly occurs in damp places such as river banks. A separate subspecies occurs in coastal sites. Angelica has a wide range of uses. Candied stems and seeds are used in confectionary, roots are a traditional herbal remedy. The root oil is used both medicinally and for flavouring liqueurs. The species is cultivated commercially in France and Scandinavia and is often in New Zealand gardens. But the common ornamental angelica in New Zealand is *A. pachycarpa* Lange.

References
Taskinen, J. and Nykanen, L. (1975). Chemical composition of Angelica root oil. Acta Chem. Scanda. (B) 29, 757–764.

Ojala, A. (1985). Variation of *Angelica archangelica spp. archangelica* in Northern Fennoscandia. 2. Phenological life strategy and reproductive output. Ann. Bot. Fennici 22(3), 183–194.

NB. For Asiatic Angelicas used in Chinese medicine see: Asahwra T., et al. (1981 et subs.). Studies on coumarins of a Chinese drug 'Quian Hu'. Part 1. Planta Medica 42, 89–96. Part 2. Jap. J. Pharmacog. 35, 331– Part 3. Planta Med. 44, 199–203 Part 4. Planta Med. 50, 117–120 Part 5. Planta Med. 50, 492–496.

Annona cherimola Mill.

Annonaceae

Common
Cherimoya, Cherimola, Custard apple (NB. *A. atemoya* and *A. reticulata* are also called custard apple. In the West Indies 'Cherimoya' is *A. reticulata*).

Notes
A subtropical fruit tree from Andean valleys of Ecuador and Peru at 1500 to 2000 m elevation. Introduced and commercially planted in northern N.Z. in the early 1980's. Possibly the hardiest member of the genus Annona which also includes:
A. x atemoya Hort. The atemoya
A. squamosa L. The sweet sop or sugar apple, apeli (Fiji).
A. glabra L. The bullock's heart or pond apple, kaitabo (Fiji). Naturalised on Viti Levu.
A. muricata L. soursop, seremaia or sarifa (Fiji). Widely planted in the islands of the South Pacific.
A. reticulata L. custard apple or bullock's heart.

References
George, A.P. and Nissen, R.J. (1985). The custard apple. Australian Horticulture 83(10), 100–111.

Sturrock, D. (1959). Fruits for southern Florida. Southeastern Printing Co., Stuart. pp 66–67.

Aralia cordata Thunb.

Araliaceae

Common
Udo (Japan), Tu–huo (China).

Synonyms
Aralia edulis Sieb. et Zucc.

Notes
A herbaceous perennial, stems 2 m or more in height. Native to Japan and north China. Cultivated for its young edible shoots harvested like asparagus. Recently introduced to N.Z.

References
Fairchild, D.G. (1903). Three new plant introductions from Japan. USDA Bulletin No. 42, Washington, pp 17–20.

Yashiroda, K. (1968). A dictionary of popular Japanese herbs and their uses today. Plants and Gardens 24(2), 51.

Figure 5: Udo plant

Arctium lappa L.

Compositae

Common
Burdock, Gobo (Japan).

Synonyms
Lappa glabra Lam., *Lappa edulis* Sieb., *Arctium bardana* Willd.

Notes
Cultivated in Japan and Hawaii for its edible root, elsewhere foliage, roots and seeds used in herbal medicine. Considered a weed in N.Z. Trails at Lincoln 1980/81 showed root production for Japanese market uneconomic due to harvesting difficulties, (Vide J. Palmer). Other trials at Tara Hills (M. Douglas) 1986/7.

References
Herklots, G.A.C. (1972). Vegetable crops in Southeast Asia. George Allen and Unwin, London. pp 461–462.

Takashima, S. et al. (1980). Vegetable crops of Japan in colour Hoikusha Publishing Co, Osaka. pp 39–40 (Text mainly Japanese).

Figure 6: Japanese gobo flowering.

Areca catechu L.

Palmae

Common
Betel nut, Areca nut, Pinang (Malay), Bunga (Philippines).

Notes
A slender palm from Southeast Asia. The nut is widely used in that area as a mild stimulant. Relatively rare and little used in the South Pacific.

References
Bavappa, K.V.A. et al. (eds) (1982). The areca nut palm. Central Plantation Crops Research Institute, Kasaragod, Kerala. 340 pp.

Farmworth, E.R. (1976). Betel nut – its composition, chemistry and uses. Science in New Guinea 4(2), 85–90.

Arnica montana L.

Compositae

Common
Arnica, Leopards bane, Wolfsbane, Mountain tobacco.

Notes
A European mountain herb much used as a vulnary in herbal medicine. The optimum habitat for production of flavenoids is said to be at 1800 m above sea level in Europe. Frequently tried as a rock garden plant in N.Z. with little success. Arnicae flos were deleted from the British Pharmaceutical Codex 1954 but are still included in the British Herbal Pharmacopoeia.

References
Tetenzi, P. (1983). Biological preconditions of aromatic and medical plant cultivation. Acta Horticulturae 132: p. 16.

British Herbal Pharmacopoeia (1983). British Herbal Medicine Association, Cowling. pp. 30–31.

Arracacia xanthorrhiza Bancr.

Umbelliferae

Common
Arracacha, Peruvian parsnip, Apio (Venezuela)

Synonyms
Arracacia esculenta DC.

Notes
A plant requiring a long growing season to produce the fleshy side roots which are the principle edible portion. Introduced to Europe in 19th century but failed. Brought to N.Z. in 1986 (vide J. Fletcher, Lincoln). Currently grown in Mexico and Puerto Rico as well as its native range of the Andes. Highly esteemed in countries where it is available. Propagation is usually by vegetative means, but plants flowered and set seed in a quarantine cell at Lincoln in the Spring of 1988. We can only speculate that this may have been stimulated by holding plants in a dry state over winter.

References
Hodge, W.H. (1954). The edible arracacha, a little known crop of the Andes. Econ. Bot. 8(3), 195–211.

National Academy of Sciences (1975). Underexploited tropical plants with promising economic value. NAS, Washington, pp 29–32.

Figure 7: Arracacha plant in flower. Lincoln quarantine, November 1988.

Artocarpus altilis (Parkins.) Fosberg

Moraceae

Common
Breadfruit, Uto (Fiji), Ulu (Hawaii and Samoa), Kuru (Cook Is.), Uru (Tahiti).

Synonyms
Artocarpus communis J.R. and G. Forst., *A. incisus* L.f., *A. incisa* Forst.f., *Rademachia incisa* Thunb., *Sitodium altile* Parkinson.

Notes
A medium sized tree indigenous to New Guinea, Western Melanesia and possibly Micronesia. Widely planted in Polynesia. Seedless fruited cultivars have been selected and are the most valued. Apart from the fruit (a syncarp), which was an important traditional food on many smaller Pacific Islands, the timber was used for canoes, the gum exuded from fresh cut stems was used for caulking and a bark cloth prepared from the bark.

References

Coenan J. and Barrau J.(1961). The breadfruit tree in Micronesia. South Pacific Quarterly Bulletin 11, 37–39 and 65–71.

Atchly J. and Cox P.A.(1985). Breadfruit fermentation in Micronesia. Economic Botany 39(3) 326–335.

Duke, J.A. (1989). Handbook of nuts. CRC Press, Boca Raton. pp 34-36.

See also: Oster G. and Oster S.(1985). The great breadfruit scheme. Natural History 94(3) 35–41.

Artocarpus heterophyllus Lam.

Moraceae

Common
Jak–fruit, Nanka (Malay and Philippines), Uto ni India (Fiji).

Synonyms
Artocarpus integrifolia L., *A. integra* Merr., *Rademachia integra* Thunb.

Notes
A medium sized tropical forest tree with large shiny leaves. The large (to 35 kg) many seeded fruit are bourne on the trunk or on large branches. The sweet aromatic flesh is used as a vegetable. The seeds are often sold separately in Southeast Asian markets and are eaten roasted. Introduced to the larger Pacific Islands but little utilised there. N.B. The smaller *A. rigidus* Bl. from Southeast Asia has also been cultivated in Fiji but has, as yet, no recorded local name.

References
Thomas, C.A. (1980). Jackfruit, *Artocarpus heterophyllus* (Moraceae), as a source of food and income. Economic Botany 34, 154–159.

Morton, Julia F. (1965). The jackfruit: culture, varieties, utilisation. Proceedings of the Florida State Horticulture Society 78, 336–444.

Asimina triloba Dunal

Annonaceae

Common
American pawpaw, Custard apple, Indian banana.

Notes
A small tree from the north and central areas of the United States with edible fruit. Not widely cultivated. Grafted cultivars are available in the USA. Trees are said to be difficult to establish and hand pollination is sometimes required. This species is rare in New Zealand.

References
Mann, J.D. (1976). *Asimina triloba* , a hardy custard–apple. Journal of the New Zealand Tree Crops Association 2(1), 23.

Anon (1983). Pawpaw: the forgotten fruit. American Horticulturalist 62(9), 1.

Burney, B. (1985). Creamy custard apple brings out the taste of the exotic. N.Z. Farmer 106(16), 100.

Creasy, R. (1986). The gardener's handbook of edible plants. Sierra Club Books, San Francisco. pp 187–190.

The California Rare Fruit Growers Handbook Vol.6 (1974), is devoted to this species. California Rare Fruit Growers, Box P, Bonsall, CA 92003, U.S.A.

Aspalathus linearis R. Dahlgr.

Leguminosae

Common
Rooibos (tea), Red bush tea.

Synonyms
Aspalathus contaminatus Druce., *A. corymbosus* E. Mey., *Borbonia pinifolia* Marl., *Psoralea linearis* Burm.

Notes
A small leguminous shrub from semi arid areas of southern Africa. The leaves are collected and dried as a herbal tea which contains no caffein. Introduced to New Zealand several times but not known to be established.

References
Cheny, R.H. and Scholtz, E. (1963). Rooibos tea, a South African contribution to world beverages. Econ. Bot. 17(2), 186–194.

Morton, J.F. (1983). Rooibos tea, a caffeinless low tanin beverage. Econ. Bot. 37(2), 164–173.

Dahlgren, R. (1968). Revision of the genus *Aspalathus* 11. The species with ericoid and pinoid leaves. 7, subgenus *Nortieria* , with remarks on rooibos tea cultivation. Botaniska Notiser 121, 165–208.

Atriplex hortensis L.

Chenopodiaceae

Common
Orache, Mountain spinach.

Notes
A hardy annual. Young plants used as a pot herb have a similar flavour to spinach. Red leaved cultivars are ornamental and are the most common form in New Zealand.

References
Organ, J. (1960). Rare vegetables for garden and table. Faber and Faber, London . pp 140–142.

Kennelly, A.G. (1952). The home vegetable garden. Department of Agriculture Bulletin 243, Wellington. pp 126–127.

Figure 8: Orache seedling.

Atropa belladonna L.

Solanaceae

Common
Deadly nightshade, Belladonna.

Notes
A shrubby herbaceous perennial from Europe. Cultivated as a source of atropine for medicine during the second World War and in further trials at Lincoln in the late 1970's. Because of its poisonous nature nightshade has been removed from most botanic gardens and is now rare in New Zealand.

References

Sievers, A.F. (1913). Individual variation in the alkaloidal contents of belladonna plants. Journal of Agricultural Research 1, 129–146.

Cairns, D. (1943). New Zealand grown botanical drugs. Pharmaceutical Journal of New Zealand. 10th November 1943. pp 17–19.

International Trade Centre (1982). Markets for selected medicinal plants. UNCTAD/GATT, Geneva. pp 83–86.

Auricularia polytricha (Mont.) Sacc.

Auriculariaceae

Common
Jew's ear fungus, Mu ehr (China).

Synonyms
Hirneola polytricha Mont.

Notes
An edible fungus growing on dead logs which is much valued by the Chinese. During the twelve year period 1872–1883 New Zealand exported close to two thousand tons of the fungus, before land clearance reduced supply to a trickle. The species is cultivated in Taiwan and elsewhere in the Far East and has a natural range over much of the South Pacific.

References

Kirk, T. (1879). On the export of fungus from New Zealand. Transactions of the N.Z. Institute 11, 454–456.

Griffin, G.W. (1884). New Zealand, her commerce and resources. Government Printer, Wellington. pp 124-128.

Church, A.H. (1890). An edible fungus of New Zealand (*Hirneola polytricha* Montagne). Kew Bulletin No. 46, 217–220.

Cheng, S. and Tu, C.C. (1978). *Auricularia spp.* In S.T. Chang and W.A. Hayes (ed.) The biology and cultivation of edible mushrooms. Academic Press, New York, pp 605–625.

See also: Parham, B.E.V. (1939). An edible fungus (*Hirneola polytricha* Mont.) Agricultural Journal, Fiji 10(1), 23–24.

Averrhoa carambola L.

Averrhoaceae
(Oxalidaceae)

Common
Carambola, Five corner, Star fruit (but not star apple – *Chrysophyllum spp.*), Kamrakh (Fiji Indian).

Notes
An evergreen tree of uncertain origin but probably from coastal Brazil. Widely cultivated in Southeast Asia, Southern China and the Philippines, but rare in the South Pacific except for Tahiti and New Caledonia. The fruit is 5 angled in cross section and glossy yellow when ripe making it an attractive component of arranged fruit bowls.

References

Sedgley, M. (1983). Oxalidaceae, in P.E. Page (ed.) Tropical tree fruits for Australia. Queensland Department of Primary Industries, Brisbane 1984. pp 125–128.

Knight, R.J. (1969). The carambola: a fruit for South Florida. Muse News (Miami) 1(3), 60–61, 75.

Azadirachta indica A.Juss.

Meliaceae

Common

Neem, Margosa tree, Nim (Hindi).

Synonyms

Melia azadirachta L., Brand.

NB; *Melia azedarach* L. without the final 'ta', is the Persian lilac or Pride of India, a common ornamental tree in many Pacific Islands and the warmer parts of New Zealand.

Notes

A small tree (to 12m.) native to dry forest areas of Southeast Asia, from India to Java. Widely cultivated and often naturalised in Pacific archipelagos. Almost every part contains contains the bitter tasting substance 'Nimbidin' and has been used in native medicine. The seed oil (margosa oil) contains 'Azadirachtin' a steroid–like compound with insecticidal properties.

References

The Wealth of India, Raw Materials. Vol.1. C.S.I.R., New Delhi 1948. pp 140–142.

Pillai, N.R. and Santhakumari, G. (1984). Effects of nimbidin on accute chronic gastro–duodenal ulcer models. Planta Medica 50 (2), 143–145, and subsequent article pp 146–148.

Schmutterer, H. et al. (1981). Natural pesticides from the neem tree. Proceedings of the First International Neem Confreence, Eschborn, West Germany, June 1980.

Vide also: Anon (1983). Work at IRRI. The Furrow 88 (1), 3.

Figure 9: Belleisle cress.

Barbarea verna (Mill.) Asch.

Cruciferae

Common

American cress, Land cress, Winter cress, Belleisle cress.

Synonyms

B. praecox (Sm.) R.Br.

Notes

A hardy winter growing annual salad crop with a flavour similar to watercress. Common in New Zealand gardens, but under utilised commercially.

References

Organ, J. (1960). Rare vegetables for garden and table. Faber and Faber, London. pp 69–70.

N.B. The Australian native winter cress is *Barbarea australis* Hook. f.

Barringtonia edulis Seem.

Barringtoniaceae

Common
Vutu kana (Fiji).

Synonyms
Barringtonia excelsa Seem., *B. seaturae* Payens.

Notes
A large forest tree endemic to Fiji producing a large edible nut. Other species of *Barringtonia* are widespread from the Indian ocean to the tropical South Pacific and the fruit of some are eaten in the Solomon Islands and the New Hebrides but only after careful preparation. *B. asiatica* (L.) Kurz. in particular is often grown as an ornamental and the fibrous heart shaped fruit are used as fishing floats.
N.B. Vutu is a common name for all *Barringtonia* spp. in Fiji. The fruit of those other than B. edulis contain saponins which are sometimes used as fish poisons.

References
Menninger, E.A. (1977). Edible nuts of the world. Horticultural Books, Stuart, Fl., p 38.

Massal, E. and Barrau, J. (1956) Food plants of the South Sea Islands. Technical Paper No. 94, South Pacific Commission, Noumea, p 32.

N.B. For range of *Barringtonia* species, see: Payens, J.P.D.W. (1968). A monograph of the genus *Barringtonia*. Blumea 15(2), 157 – 263.

Figure 10: Young plant of Ceylon spinach.

Winters, H.F. (1963). Ceylon spinach, *Basella rubra*. Econ. Bot. 17, 195–199.

Basella alba L.

Basellaceae

Common
Ceylon spinach, Malabar nightshade, Gendola (Malay), Poi (Fiji)

Synonyms
Basella rubra L., *B. cordifolia* Lam.

Notes
Climbing succulent vine, widely grown as a 'spinach' in the tropics. In the Pacific the red leafed variety is more frequently grown as an ornamental.

References
Martin, F.W. (1979). Edible leaves of the tropics. 2nd edn. Golden Press, New York. pp 25–27.

Benincasa hispida (Thunb.) Cogn.

Cucurbitaceae

Common
Wax gourd, Chinese preserving melon, Kondol, (Philippines), Ashgourd, Togwa (Japan), Dungwa (Hawaii)

Synonyms
B. cerifera Savi.

Notes
A gourd from East Asia, cultivated in the South Pacific and may have been confused with *Lagenaria* in some early records. Can be used as a rootstock for melons.

References

Sundararajan S and Mutrukrishnan (1982). A high yielding ash gourd. Indian Horticulture 27(3), 15.

Herklots, G.A.C. (1972). Vegetables in Southeast Asia. Allen & Unwin, London. pp 307–313.

Bixa orellana L.

Bixaceae

Common

Annatto, Annotto, Roucou (West Indies), Loa (Samoa), Kesa or Qisa (Fiji). Sometimes called Polynesian lipstick.

Notes

A small evergreen shrub from tropical America. The seeds have an dark orange outer covering to the testa which produces an edible dye. Widely grown in the Pacific islands as an ornamental or hedge plant.

References

Ingram, J.S. and Francis, B.J. (1969). The annatto tree – a guide to its occurence, cultivation, preparation and uses. Tropical Science 11(2), 97–102.

Parham, W.L. (1938). Annatto. Agricultural Journal, Fiji 9(2), 24.

Boehmeria nivea (L.) Gaudich.

Urticaceae

Common

Ramie, China grass, Rhea, Chinese hemp, Lami (Samoa). N.B: Solidenia fibre, resembling wool, is derived from ramie by a special process.

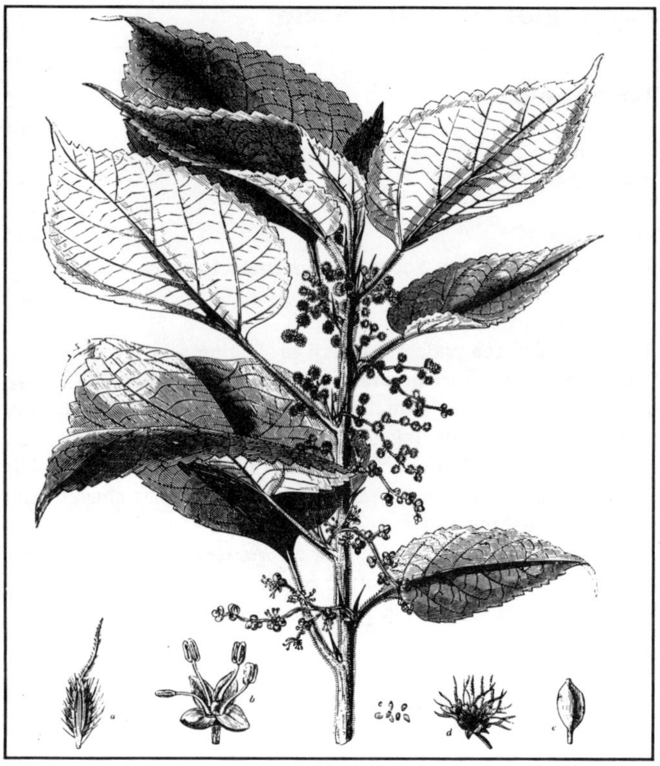

Figure 11: Ramie

Notes

Cultivated for its long, tough silky fibres derived from the inner bark (cf. *Linum*), in China and occasionally elsewhere in warm temperate regions. In the tropics it is replaced by *B. nivea var tenacissima* (Roxb.) Miq. The utility of both varieties is limited by the difficulty of fibre preparation. Trials have been conducted in New Zealand and several Pacific Islands but the crop has never been successfully established. The most recent N.Z. trial was probably that at Rukuhia Research Station in the early 1960's. A shrubby species, *Boehmeria virgata* (Forst. f.) Guillemin, koldo or rabi in Fijian, is native to Fiji and some other Pacific Islands. It has many local medicinal uses.

References

Willimot, S.G. (1954). Ramie fibre its cultivation and development. World Crops 6(10), 405–408.

Kirby, R.H. (1963). Vegetable fibres. Loenard Hill, London. pp 148–179.

Morgan, W. (1887). Rhea or China grass. The N.Z. Country Journal 11 (1), 55–59.

Bell, W.T. (1898). On the cultivation of Rhea or Ramie–plant (*Boehmeria nivea* Hook. & Arn.) N.Z. Department of Agriculture Leaflet for Farmers No. 42, 4p. 1st issue Wellington 1898 reprinted 1900.

Brassica campestris L. *ssp. pekinensis* (Lour.) Olssen

Cruciferae

Common
Chinese cabbage, Celery cabbage, Wong Bok, Pe–tsai

Synonyms
Brassica chinensis L., *B. orientalis* Thunb.

Notes
The principal green vegetable of Northern China. *B. campestris ssp. chinensis* (Rupr.) Olsson, the 'pak choi', predominates in southern China. Both are widely cultivated in the tropics and warm temperate regions. Except in China the common names do not distinguish between the subspecies.

References
Palmer J. (1981). The Chinese brassicas. Annual Journal of Royal N.Z. Institute of Horticulture No. 9, 72–74.

Talekar, N.S. and Griggs T.D. (1981) eds. Chinese cabbage. Proceedings of the first International Symposium. AVRDC, Shanhua. 489 pp.

Brassica carinata A. Br.

Cruciferae

Common
Ethiopian mustard.

Notes
B. carinata is an amphidiploid (B. oleracea x B. nigra), not known as a wild species. It is cultivated as a leaf vegetable particularly in Ethiopia and East Africa, and as an oil seed crop.

References
Stephens, T.S. et al. (1970). Quality of *Brassica carinata* as a green leafy vegetable. J. Amer. Soc. Hortic. Sci. 95, 3–5.

Katiyar, R.K. et al. (1986). Evolution of *Brassica carinata* as a new oilseed crop for India. Experimental Agriculture 22(1), 67–70.

Brassica juncea (L.) Czerniac

Cruciferae

Common
Brown mustard, Mustard rape.

Notes
Grown as a leaf vegetable in China, the Himalayan region and to a limited extent in the U.S.A. Also as an oil seed crop in India and elsewhere and a weed of some New Zealand roadsides. NB, it should not be confused with the Indian 'sarson' which is *B. campestris ssp oleifera.* Brown mustard plants usually grow slowly. They have a high tolerance of heat and drought but a low tolerance of cold.

References
Chen, Shi–ru (1982). The origin and differentiation of mustard varieties in China. Cruciferae Newsletter 7, 710.

Hemingway, J.G. et al. (1961). Volatile mustard oils in Brassica juncea seeds. Nature 192, 993.

Li, C.W. (1980). Classification and evolution of mustard crops in China. Cruciferae Newsletter 5, 33–35.

Brassica rapa L. (Ravo group)

Cruciferae

Common
Broccoli raab, Italian flowering turnip.

Synonyms
Brassica campestris L.

Notes
An uncommon vegetable in N.Z. Both spring and autumn flowering cultivars are available in the U.S.A. The stem does not become swollen as in the common turnip.

References
Johnson, W.B. (1977). Broccoli raab in New Jersey. American Vegetable Grower 25(9), 14.

Anon (1986). Broccoli raab. Organic Gardening 33(9), 80.

Figure 12: Broccoli raab plant.

Broussonetia papyrifera (L.) Vent.

Moraceae

Common
Paper mulberry, Aute (Maori)*, U'a (Samoa), Malo or Ai Maisi (Fiji), Hiapo (Tonga and Niue), Kaji–no–ki (Japan).
*N.B. Aute in modern Tahiti refers to *Hibiscus rosa–sinensis*.

Synonyms
Morus papyrifera L.

Notes
A small deciduous tree native to China and Japan where the bark was traditionally used for paper. Probably introduced to the South Pacific (including N.Z.) by Polynesian voyagers and at one time cultivated throughout the area as a source of 'tapa' bark, cloth. It was rare in New Zealand at the time of Cook's first visit and apparently died out altogether shortly thereafter. The paper mulberry is now rare or absent from many Pacific Islands, including Tahiti and Niue, where it was once common. The tree is dioecious and hardy throughout lowland New Zealand.

References
Sykes, W.R. (1969). *Broussonetia papyrifera*. An unusual case of sex reversion. Journal of the Royal N.Z. Institute of Horticulture. New Series 1(2), 63–67.

Seemann, B. (1862). Viti : an account of a government mission to the Vitian or Fijian Islands in the years 1860–61. Macmillan, Cambridge, pp 348–351 (for detailed account of the manufacture of tapa bark cloth. This note was largely reprinted in his Flora Vitiensis of 1868, p 246).

Degener,O. (1945). Plants of Hawaii National Park illustrative of plants and customs of the South Seas. Edwards Brothers, Ann Arbor. pp 132–142.

N.B. For Maori use of *Broussonetia* vide:

Best E. (1976). Maori agriculture. 2nd edn. Government Printer, Wellington. pp 29–31 and 277–278.

See also: Kennedy, R. (1934). Bark cloth in Indonesia. Journal of the Polynesian Society 43(4), 220–243.

Cajanus cajan (L.) Millsp.

Leguminosae

Common
Pigeon pea, Ahar (Hindi), Kachang kayu (Malay), Juto (Japan).

Synonyms
Cajanus indicus Spreng., *C. bicolor* DC.

Notes
A shrubby perennial legume producing a high value seed. Widely grown particularly by people of Indian origin from East Africa to Fiji. Recently commercialised in Queensland. Dwarf early maturing cultivars were grown in trials at Otara in 1974–5 and more recently by MAF in Central Otago.

References
Morton, J.F. (1976). The pigeon pea, a high protein tropical bush legume. Hort. Science 11, 11–19.

Gooding, H.J. (1962). The agronomic aspects of pigeon peas. Field Crop Abstracts 15, 1–4.

See also: Thomas P.I. and Leatio'o (1985). Pigeon pea: A versatile recent introduction to Western Samoa. Alafua 10(1), 1–7.

Figure 13: Pigeon pea at Otara Research Station 1975.

Figure 14: Japanese tea plant at Riwaka Research Station.

Camellia sinensis (L.) Kuntze.

Theaceae

Common
Tea, Chah (Hindi).

Synonyms
Camellia theifera Griff., *C.thea* Link., *Thea chinensis* Sims.

Notes
Camellia sinensis is a small shrub from southern China, cultivated there and in Japan. Recently introduced to Nelson Province on a commercial basis. The larger leafed *C. sinensis var. assamica* (Mort.) Kitam. is cultivated in India and East Africa. This forms a small tree when not heavily pruned. Var. assamica was planted on the West Coast of the South Island in the late 1960's and some plants are still growing there (vide Walton 1971).

References

i) Hobman, F.R. (1985). Effects of plant population and time of plucking on the production and profitability of tea in Australia. Tropical Agriculture (Trinidad) 62(3), 193–198

Seshadri R., et al. (1986). Utilisation of biproducts of the tea plant : A review. Tropical Agriculture (Trin.) 63(1), 2–6.

ii) N.Z. References.

Anon (1988). Green tea growing at Motueka. Growing Today. December/January 1988, p 10.

Walton, T. (1971). Tea trials on the West Coast. N.Z. Journal of Agriculture July 1971, p 8.

'D.M.' (1887). Tea growing in New Zealand. The N.Z. Country Journal 11 (6), 490–495.

iii) Other teas:

Puerh tea. Kew Bulletin 1892, pp 118–120.
Paraguay tea. (syn. mate tea – *Ilex paraguariensis*). Kew Bulletin 1892, pp 132–137.
Faham tea. Kew Bulletin 1892, pp 181–183.
Lao tea. Kew Bulletin 1892, pp 219–222.
Mauritius tea. Kew Bulletin 1892, pp 234–238.
Leppett tea. Kew Bulletin 1892, pp 10–13.
White tea of Persia. Kew Bulletin 1892, pp 157–158.

Cananga odorata (Lam.) Hook. f. and Thoms.

Annonaceae

Common
Motoi (Tahiti), Mosooi (Samoa), Makosoi (Fiji), Ylang–ylang (Malesia).

Synonyms
Uvaria odorata Lam., *Canangium odoratum* Baill.

Notes
An evergreen tree (to 20m). Native to tropical Asia, cultivated and naturalised in many Pacific Islands. The flowers which occur throughout most of the year are used for scenting coconut oil and in necklaces.

References
Uhe, G. (1974). Wayside plants of the South Pacific. Stocton House, Auckland. pp 129–130.

Burkill, I,H. (1935). Dictionary of the economic products of the Malay Peninsula. Crown Agents, London. Vol 1, pp 426–429.

Petard, P. (1986). Plants utiles de Polynesie. Haere po no Tahiti, p 162.

Canarium album (Lour.) Raeuschel and *Canarium pimela*

Burseraceae

Common
Chinese olives, white and black respectively.

Synonyms
1) *Pimela alba* Lour. and *Canarium tonkinense* Engler.
2) *Canarium pimela* Konig. and *C. nigra* (Lour.) Engler.

Notes
Widely cultivated in southern China, Taiwan, and Indo China. The fruit is usually pickled but both species are also used medicinally in the dried state.

References
Little up–to–date information is available on these species.

Porterfield, W.M. (1951). The principal Chinese vegetable foods and food plants of Chinatown markets. Economic Botany, 5 (1), 11–17.

Chun, Woon–Yeung. (1921). Chinese economic trees. Commercial Press, Shanghai. pp 200–201.

Hance, H.F. (1871). On the so called 'olives' of southern China. Journal of Botany 9, 38–40.

Walker, E.H. (1954). Important trees of the Ryukyu Islands Spec. Bull. US Civ. Adm. Ryukyu Is No. 3. 350 pp. (Not sighted).

Canarium spp. (other than Chinese olives).

Burseraceae

Common
C. vulgare and *C. indicum* are often called Java almonds or Kenari. *C. luzonicum* and *C. ovatum* are Pili nuts.

Notes
A great many Canarium species are harvested for their seed (nuts) where they occur naturally. Only those species with superior nuts are cultivated. In the Philippines, *Canarium luzonicum* is an important source of resin – Manila elemi.
Some 40 species of Canarium occur in South and East Asia through to the South Pacific. The nomenclature of the genus is confused. Vide: Leenhouts, P.W. (1959) Revision of the Burseraceae of the Malaysian area, *Canarium* Stikm. Blumea 9(2), 275–471.
More important economic species include:
C. harveyi Seem. occurs naturally as far east in the Pacific as Niue and Samoa.
C. vulgare Leenhouts (syn. *C. commune* L.), from Bali and Lesser Sunda Islands, but widely cultivated in Indonesia.
C. indicum L. (syn. *C. vulgare* Rumph. & *C. moluccanum* Blume.) from the Moluccas through New Guinea to the Solomons and New Hebrides.
C. luzonicum (Bl.) A. Gray (syn. *C. commune* Fernandez Villar and *C. ticandrum* Engler) from the Philippines.
C. ovatum Engler is also from the Philippines but with an extended range through parts of Malaysia and Melanesia.

References
Menninger, E.A. (1977). Edible nuts of the world. Horticultural Books, Stuart. pp 25–28.

Duke, J.A. (1989). Handbook of nuts. CRC Press, Boca Raton. – *C. indicum* , pp 64–65. *C. ovatum* , pp 67–68.

Brown, W.H. (1921). Minor products of Philippine forests. Bureau of Forestry Bulletin No. 22, Manila. pp 40–49 (resins).

Burkill, I.H. (1935). Dictionary of the economic products of the Malay Peninsula. Crown Agents, London. pp 424–432.

Allarquez, A.H. (1982). Pili management for resin and nut production. Canopy International 8(4), 14-16.

Canavalia ensiformis (L.) DC.

Leguminosae

Common
Jack bean, Horse bean, Sabre bean, Natamane (Japan), Jenja (New Guinea).

Synonyms
Dolichos ensiformis L.

Notes
Jack beans are quick growing, large podded, climbing legumes adapted to marginal soils in the Tropics. They are native to Central America but cultivated worldwide.
NB Sword bean is the closely related *C. gladiata* (Jacq.) DC. widely cultivated in India and Africa.

References
National Academy of Sciences. Tropical legumes : resources for the future. Washington, DC, 1979. pp 54–59.

Purseglove, J.W. (1968). Tropical crops : Dicotyledons Vol. 1. Longman, London. pp 242–244.

Canna edulis Ker Gawler

Cannaceae

Common
Queensland arrowroot*, Edible canna, Achira (Bolivia and Peru).

Synonyms
Canna esculenta Lodd., *C. rubricaulis* Link.

Notes
Subtropical herb, unable to stand prolonged exposure to wind. Still grown in Queensland for the production of edible starch. Introduced to many areas of Melanesia but not grown commercially there. Cannot compete with *Maranta arundinacea* in West Indies as a source of edible starch on world markets (?)
N.B.* For arrowroot see *Maranta arundinacea* .

References
Gade, D.W. (1966). Achira, the edible canna, its cultivation and use in the Peruvian Andes. Econ. Bot. 20(4), 407–415.

Ufer, M. (1972). *Canna edulis,* a neglected rootcrop. Tropical root and tuber crop newsletter (5), 32–34.

Capparis spinosa L.

Capparidaceae

Common
Caper bush, Kabra (Hindi).

Notes
A spiny deciduous shrub from the Mediterranean region through to Northwestern India. Capers are made from the flower buds. Rarely grown in the Pacific region. The species found in many Pacific islands is *C. cordifolia* Lam. which could be native to the region (Vide; W.R. Sykes. (1970). Contributions to the Flora of Niue. DSIR Bull 200, p56). The 'caper bush' of Australia is *C. nummularia* DC., but this and some 20 other native Australian species of Capparis are not much utilised and only the fruit were eaten by the Aboriginees.

References
Barbara, G. and Di Lorenzo, R. (1984). The caper culture in Italy. Acta Horticulturae 144, 167–171.

Orphanos, P.I. (1983). Germination of caper seeds. J. Hort. Sci. 58(2), 267–270.

Creasy, R. (1986). The gardener's handbook of edible plants. Sierra Club Books, San Francisco. pp 71–73.

Carica papaya L.

Caricaceae

Common
Pawpaw, Papaya, Ninita (Cook Is), Lesi (Tonga), Oleti or Weleti (Fiji).

Notes
A small unbranched tree, with large palmate leaves and dioecious, rarely hermaphrodite flowers. Probably originated in Central America but is now cultivated throughout the tropics.

References
Yeo, W. et al (1970). Papayas in Hawaii. Hawaii Coop Ext. Serv. Circ. 436, 57 pp. (Reprinted 1979).

Agnew, G.W. (1968). Growing papaws in Queensland. Queensl. Agric. J. 94, 24–36.

Carica pubescens Lenne. et Koch.

Caricaceae

Common
Mountain pawpaw, Chamburo (Ecuador).

Synonyms
Carica candamarcensis Hook. f., *Papaya pubescens* Kuntze, *Carica chiriquensis* Woodson.

Notes
Native to mountain areas between 1500 and 3000 m from Panama to Bolivia. Much hardier than the true pawpaw and can withstand light frosts. Cultivated in Chile for processing. Widely grown in gardens of northern New Zealand but fruit seldom utilised. Plants are monoecious. Plants which initially only produce male flowers normally produce some small fruits when plants mature.

References
MacKenzie, R.A. and Strachan, G. (1980). Possibilities for processing mountain papayas in N.Z. Food Technology in New Zealand 15(8), 13 and 19.

Pearson, H. (1980). The pawpaw. 16 pp. cyclostyled report. N.Z. Tree Crops Association, Auckland. *C. pubescens* is discussed both as *C. pubescens* and *C. candamarcensis*, which are included as separate species.

Carica spp.

Caricaceae

Notes
Other Carica species from the high Andes recently introduced into N.Z. include:
Carica monoica Desf. syn. *C. boliviana* a short lived species from Ecuador and eastern Peru, at between 600 and 1700 m elevation. Commonly called 'col de mont', literally mountain cabbage, because of its short stature (1 to 3 metres).The fruit is small (7.5 x 5 cm), fragrant and said to be of good flavour especially when cooked. Basal shoots are produced freely.
C. stipulata Badillo, the siglalon or toronche comes from a restricted area of southern Ecuador at between 1600 and 2450 m. Plants are monoecious and the fruit are small, but carried all the way up the stem. The fruit is said to be variable in quality.
C. parviflora A.DC. has red flowers and is usually grown as an ornamental.

Carica x heilbornii Badillo

Caricaceae

Common
Babaco.

Synonyms
Carica pentagona Heilborn.

Notes
A naturally occuring hybrid of *C. pubescens x C. stipulata* from Ecuador. Introduced to New Zealand in the late 1970's and commercially planted in 1981. Fruit are parthenocarpic and sterile. Propagation is vegetative.

References
Endt, D. (1982). Babaco progress 1982. The Orchardist of New Zealand 55(6), 170, 172 & 174.

Endt, D. (1986). Growing interest in babaco worldwide. The Orchardist of New Zealand, 59(8), 307 & 310.

N.B. Badillo, V.M. (1971). Monografia de la familia Caricaceae. Publicado por la Asociation de Profesores, Universidad Central de Venezuela, Maracay, 209 pp.

Badillo lists 5 nothomorphs (hybrid varieties) occuring in *C. x heilbornii*. The N.Z. babaco is *C. x heilbornii nm pentagona* (Heilborn) Badillo.

Carthamus tinctorius L.

Compositae

Common
Safflower.

Synonyms
C. dentatus Blanco.

Notes
Widely cultivated for a dye obtained from the flowers, as an oil seed and for medicine in the Far East. In northern India, oil is converted by a special process into 'roghan', a jelly like substance used in the preparation of wax cloth for linoleum manufacture. Cultivation in New Zealand restricted by susceptibility to 'Botrytis'.

References
Beech, D.F. (1969). Safflower. Field Crop Absracts, 22, 107–119.

Manning, S.H. et al. (1974). Investigations into the development of oil seed crops in N.Z. Proceedings, Agronomy Society of N.Z. 4, 19–23.

See also: Painter, G. and Power, E. (1982). A garden of old fashioned and unusual herbs. Hodder and Stoughton, Auckland. p 68.

Aglink FPP662. Sunflower and safflower for seed and oil. MAF, Wellington 1982.

Carum carvi L.

Umbelliferae

Common
Caraway.*

Notes
A biennial herb native to southern Europe and Asia Minor; naturalised in North America. Most cultivars require some 15 months to maturity but annual and tetraploid cultivars have been reported. The seeds (mericarps) are used as flavouring agents, in liqueurs (e.g. kummel) and in medicine. An occasional crop in New Zealand, but no trial work reported. Caraway is sensitive to frost, plant competition and mechanical injury.
*NB Black caraway is *Nigella sativa,* mostly produced in the Middle East.

References
P. Maghami (1984). Blond caraway (*Carum carvi* L.). Acta Horticulturae 144, 21–24.

Simon, J.E. et al.(1984). Herbs. An indexed bibliography 1971–1980. Archon Books, Hamden, Conn. pp 17–18.

Anon (1885). Caraway seed. The N.Z. Country Journal 9(4), 311.

Casimiroa edulis Llave

Rutaceae

Common
White sapote, Casimiroa.

Notes
A fruit tree from Central America and the highlands of Mexico. Very drought resistant. The flowers are dioecious or occasionally hermaphrodite. Planting recently commenced on a small scale in New Zealand. More widely planted in Australia from Sydney to the Atherton Tableland.

References
George, A.P. (1986). Casimiroa; white sapote. Australian Horticulture 87(7), 38 and 44.

Dawes S N. (1969). The casimiroa, a promising subtropical fruit. Orchardist of New Zealand 42(7), 224–225,227.

Gardner P.(1987). Casimiroa action group. Growing Today 4(5),44.

Cassia spp.

Leguminosae

Notes
Several species of Cassia are used in medicine:
Cassia acutifolia Del., Alexandrian senna – foliage and pods official.
Cassia obtusifolia L., seed used in Chinese medicine.
Cassia tora L., (syn. *Senna tora* (L.) Roxb.) seed used in Chinese medicine.
The last species is a weed on some Pacific Islands. The two others are more adapted to the arid tropics.
N.B. Cassia bark is derived from the Chinese tree *Cinnamomum cassia.*

References
ITC (1982). Markets for selected medicinal plants and their derivatives. UNCTAD/GATT, Geneva. pp 159–167.

Gupta, R. (1984). Senna cultivation in India. Indian Horticulture 28(4), 19–21, 48.

Vide also: Divar, H.A. (1984). Kawal, meat substitute from fermented *Cassia obtusifolia* leaves. Economic Botany 38(3), 342–349.

Catha edulis (Vahl) Forsk. ex Endl.

Celastraceae

Common
Chat, Kaht, Qat, Cafta, Arabian tea.

Notes
A small shrub from Arabia, Ethiopia and East Africa. The leaves produce a stimulating beverage and are chewed by the Arabs as a narcotic. Usually sold fresh as bundles of twigs. An occasional ornamental in New Zealand gardens.

References
Krikarian, A.D. (1973). Chat, coffee's rival from Harrar, Ethiopia. 1) Botany, cultivation and use. Economic Botany 27(4), 353–377. 2) Chemical composition. ibid. 378–389.

U.N. Bulletin on Narcotics 32(3), (1980). Special issue on khat.

Catharanthus roseus (L.) G. Don

Apocynaceae

Common
Madagascar periwinkle, Old maid, Church flower, Myrtle, Magdalena.

Synonyms
Vinca rosea L., *Ammocallis rosea* Small, *Lochnera rosea* Reichb. f.

Notes
Probably native to the West Indies but now spread widely through most tropical regions as an ornamental and in the tropical Pacific as a weed. It is a constituent of herbal medicines from Cuba to Southeast Asia. Periwinkle alkaloids are prescribed for leukemia and in cancer chemotherapy. As a medicinal crop periwinkle has been cultivated in Africa, India, Hungary and Australia but its temperature requirements are probably too high for New Zealand.

References
Taylor, W.I. and Farnsworth, N.R. (1973), eds. The vinca alkaloids; botany chemistry and pharmacology. M. Dekker, New York. 357 pp.

Datta, S.C. (1982). Cultivation of *Catharanthus roseus* in West Bengal. In C.K. Atal and B.M. Kapur (eds.) Cultivation and utilisation of medicinal plants CSIR, Jammu–Tawi (India). pp 279–283.

Celosia argentea L.

Amaranthaceae

Common
Cockscomb, Sokoyokoto (Africa), Celosia.

Notes
Celosia is an important leaf vegetable in tropical Africa. In Southeast Asia it is often considered to be another variety of Amaranthus but differs from this genus in having a C3 cycle photosynthetic pathway. The species probably originates from India but is little utilised there or in the Pacific.

References
Khoshoo, T.N. and Pal, N. (1973). The probable origin and relationships of garden cockscomb. Botanical Journal of Linnean Society 66, 127–141.

Pal, M. and Khoshoo, T.N. (1983). Utilisation of resources of *Celosia argentea*. Plant Genetic Resources Newsletter 53, 22–24.

Martin, F.W. and Telek, L. (1979). Vegetables for the hot humid tropics, 6. Amaranth and Celosia. USDA, New Orleans, LA. 21 pp.

Cephaelis ipecacuanha (Stokes) Baill.

Rubiaceae

Common
Ipecac, Hippo, Goldenroot.

Synonyms
Psychotria ipecacuanha Stokes., *Uragoga ipecacuanha* Baill.

Notes
A herbaceous plant with a corky underground stem and long lateral roots from humid forest areas of Brazil and Bolivia. It has been used in Europe since the 17th century as a treatment for dysentery and as an emetic, for which use it is still included in most pharmacopoeias today. Ipecac is cultivated in Brazil and India. Pre-war it was grown in Malaysia and included in trials in Pacific Islands.

References
Fisher, H.H. (1973). Origin and uses of ipecac. Economic Botany 27, 231–234.

Chattejee, S.K. et al. (1982). Cultivation and utilisation of ipecac in West Bengal. In C.K. Atal and B. Kapur (eds) Cultivation and utilisation of aromatic and medicinal plants. CSIR, Jamma–Tawi (India), pp 295–301.

N.B. Striated or black ipecac from Colombia is derived from a related species. *C. emetica* Pers. (Syn. *Uragoga emetica* Baill.). Cartegena, Costa Rica, Nicaragua or Panama ipecac is from *C. acuminata* Karsten.

Ceratonia siliqua L.

Leguminosae

Common
Carob, St John's bread, Al kharoubah (Arabic), Locust bean.

Notes
A deciduous tree from the Eastern Mediterranean with edible sugary pods. A gum resembling tragacanth is also produced from the pods, and the seeds used in the preparation of a diabetic flour. A tree for warm arid sites in New Zealand. Much more widely planted in Australia.

References
Winer, N. (1980). The potential of the carob. International Tree Crops Journal 1(1), 15–26.

Ebenshade, H. and Wilson, G. (1986). Growing carobs in Australia. Goddard and Dobson, Box Hill. 113 pp.

Chaerophyllum bulbosum L.

Umbelliferae

Common
Bulbous chervil.

Synonyms
Scandix bulbosa Roth, *Selinum bulbosum* E.H.L. Krause, *Myrrhis bulbosa* Sprengel.

Figure 15: Bulbous chervil seedling.

Notes
A small starchy rooted biennial herb from central Europe extending as far west as Germany, but not introduced to France and Great Britain until the middle of the 19th century. The root resembles a short stumpy carrot, yellow internally, brown or purplish skinned. Now rarely cultivated because the seed needs vernalisation. Introduced to N.Z. from Vilmorin in France, 1980.

References
Organ, J. (1960). Rare vegetables for garden and table. Faber and Faber, London. pp 87–88.

Peron, J–Y. (1983). Le cerfeuil tubereaux. Revue Horticole, 233, 32–33.

Hegi, G. (1926). Illustrierte Flora von Mittel Europa. Munchen, Vol. 5(2), pp 998–1001.

Chamomilla recutita (L.) Rausch

Compositae

Common
Chamomile, German chamomile.

Synonyms
Matricaria chamomilla L., *Matricaria suaveolens* (Pursh) Rydb., *Matricaria recutita* L.

Notes
A quick growing annual herb from Europe and Western Asia. Small areas grown in Canterbury since 1980 for the production of flower heads used in chamomile tea.

References
Franz Ch. et al. (1986). Influence of ecological factors on yield and essential oil of camomila. Acta Horticulturae 188, 157–162.

Singh, A. (1982). Cultivation of *Matricaria chamomilla* . In; C.K. Atal and B.M. Kapur (eds). Cultivation and utilisation of aromatic and medicinal plants. CSIR, Jammu–Tawi, (India). pp 553–658.

N.B. Roman chamomile is *Chamaemelum nobile* (L.) All. (syn. *Anthemis nobilis* L.), also used in herbal medicine but seldom cultivated on a commercial scale.

Figure 16: Chamomile plant at harvesting stage.

Chenopodium bonus–henricus L.

Chenopodiaceae

Common
Good King Henry, Allgood, Lincolnshire asparagus.

Notes
An ancient perennial European vegetable crop used as a spinach substitute. Naturalised on some roadsides of Canterbury and Otago and cultivated in a few gardens.

References
Saunders, P. (1985). Good King Henry. Garden 110(12), 579–580.

Barney, Sarah (1984). When goosefoots stepped into the New World. New Scientist No. 1430, p 24.

N.B. Plants of the *Chenopodium album* complex are used as pot herbs in the Himalayas.

Chenopodium quinoa Willd.

Chenopodiaceae

Common
Quinoa, Quinua.

Notes
An Andean cereal crop grown in Peru, Bolivia and Chile. The small seeds contain bitter tasting saponins in the testa which are removed by washing and cooking. Several accessions were introduced to Canterbury in 1980. Trials were discontinued due to inability to control weeds, but plants seemed better adapted than amaranths. Recently (1988/9) further plots have been planted at Lincoln College (vide R. Crowder). N.B. *Chenopodium pallidicaule* Aellen (syn. *C. canihua* Cook), the canihua is a smaller growing species cultivated at even higher altitudes in the Andes. *C. berlandieri* is thought to have been domesticated in the Mexican highlands and is still cultivated in home gardens for the edible immature inflorescences called 'huauzontle'.

References
Weber, E.J. (1978). The Inca's ancient answer to food shortage. Nature 272, 486.

Wilson, H.D. (1983). Quinua: significant past – questionable future. The Herbarist 49, 115–120.

Figure 17: Quinua.

Chrysophyllum cainito L.

Sapotaceae

Common
Star apple, Caimito, Pipi o eva (Samoa).

Synonyms
Achras caimito Ruiz. et Pavon.

Notes
An ornamental fruit tree (to 10m). The oval dark green leaves have a brown tomentum on the underside. The globular purplish skinned fruit is sectioned internally in the form of a star. Widespread but uncommon in the South Pacific.

References
Parham, J,W. (1948). The Botanical Gardens, Suva. Agricultural Journal, Fiji 19(2), 102.

Morton, Julia M. (1987). Fruit of warm climates. Creative Resource Systems, Winterville. pp 408–410.

Cinnamomum camphora (L.) J.S.Presl

Lauraceae

Common
Camphor, Camphor laurel.

Notes
A large evergreen tree native to China, Taiwan and Japan. Camphor is distilled from old wood cut into chips. But in cultivation trees may be coppiced and camphor distilled from shoots. Camphor has been widely planted in warm temperate regions and is grown in Australia, especially around Sydney, and occasionally in northern New Zealand. Synthetic camphor from turpentine is now a serious competitor of the natural product.

References
Youngman, Barbara J. (1952). Professor Naonari Hirota's work on camphor trees. Kew Bulletin 1952, 61–65.

King, F.A. (1958). From camphor chips to medicinal preparations. The Australasian Journal of Pharmacy 39, 1160.

See also: Kirk, T. (1881) Camphor. Trans. N.Z. Institute 13, 136.

Cinnamomum zelandicum Bl.

Lauraceae

Common
Cinnamon, Kaloni (Fiji), Tigamoni (Samoa).

Notes
A medium sized evergreen tree (to 15m.) probably indigenous to Sri Lanka. Cinnamon is the dried bark from young shoots. The main areas of production include Sri Lanka, Seychelles and the Malagasy Republic. Cinnamon has been introduced to most Pacific Islands but is nowhere commercially exploited.

References
Purseglove, J,W, et al. (1981). Spices. Longman, London Vol. 1, pp 100–173.

Samarawira, I,St E. (1964). Cinnamon. World Crops 16, 45–48.

Citrullus colocynthus (L.) Schrad.

Cucurbitaceae

Common
Colocynth.

Synonyms
Colocynthus vulgaris Schrad.

Notes
A bitter gourd from the Mediterranean area and through to Northwestern India. Widely used in medicine as a cathartic (Vide: British Pharmaceutical Codex, 1963 pp 206–208). Cultivated in Spain, Cyprus and Turkey.

References
Anon (1950). The wealth of India – Raw Materials Vol. 2. CSIR, New Delhi. pp 185–186.

Cnidoscolus chayamansa McVaugh

Euphorbiaceae

Common
Chaya, Tree spinach.

Notes
A shrub, 3 to 5 m tall, native to Honduras and Mexico. Usually propagated by cuttings. The shoots and young leaves are edible but are hairy and chewy in texture. Hydrocyanic glucosides present are removed in cooking.

References
National Academy of Sciences (1975). Underexploited tropical plants with promising economic value. Washington. pp 45–48.

Vietmeyer, N. (1985). In praise of shrubs. Ceres 18(2), 28–32.

Coccinia grandis (L.) Voigt

Cucurbitaceae

Common
Scarlet gourd, Ivy gourd, Kundru (Hindi).

Synonyms
Bryonia grandis L., *Coccinia cordifolia* Cogn.

Notes
A scandent herb from tropical Asia, Malesia and northern Australia. Introduced to many Pacific Islands and established as a weed in Fiji where the fruits are eaten by the Indian population. The ovoid– elliptic fruit ripen to a bright red colour. In India the fruit is used both for food and medicine. It is a potential remedy for diabetes.

References
Ramachandran, K. and Subramanian, B. (1983). Scarlet gourd, Coccinea grandis, little known tropical drug plant. Economic Botany 37 (4), 380–383.

The Wealth of India, Raw Materials Vol. 2. CSIR, Delhi 1950. p 257.

Coix lacryma–jobi L.

Gramineae

Common
Job's tears, Adlai (Philippines), Sila (Fiji), Tangatanga (Polynesia).

Synonyms
C. agrostis Lour., *C. arundinacea* Lam., *C. exaltata* Jacq., *C. lachrima* L., *C. ovata* Stok., *C. pendula* Salisb., *C. stigmatosa* Koch., *Lithagrostis lachryma–jobi* Gaertn.

Notes
A perennial grass species occuring over a wide area of south and east Asia and in many Pacific Islands. There are several botanical varieties, including var *ma–yuen* Stapf which is grown as an annual and used in Chinese medicine. The seeds are used as a minor cereal in many areas but not in the Pacific(?). In New Zealand perennial accessions probably originating from Pacific Islands, are grown as garden ornamentals. It is hardy as far south as Akaroa in sheltered sites.

References
Arora, R.K. (1977). Job's tears, a minor food and fodder crop of North-east India. Economic Botany 31, 358–366.

Jain, S.K. and Banerjee, D.K. (1974). Preliminary observations on the ethnobotany of the genus Coix. Economic Botany 28(1), 38–42.

Coleus aromaticus Benth.

Labiatae

Common
Country borage (India), Indian borage, Spanish thyme, Oregano (West Indies), Bangun bangun (Malay), Suganda (Philippines), Paharchur (Hindi).

Synonyms
Coleus amboinicus Lour., *C. carnosus* Hook., *C. crassifolia* Benth., *C. suganda* Blanco, *Plectranthus amboinicus* Spreng., *P. aromaticus* Roxb.

Notes
An aromatic tropical herb with thick fleshy leaves on lax stems up to 1 metre in length. Origin unknown, but naturalised in parts of India and widely grown as an occasional herb or medicinal plant throughout the tropics and sub tropics. Usually a pot plant in N.Z.

References
Merrill, E.D. (1937). *Coleus amboinicus* , suganda. Addisonia 20, 11-12.

Herklots, G.A.C. (1972). Vegetables in Southeast Asia. George Allen and Unwin, London. pp 479–481.

Coleus parviflorus Benth.

Labiatae

Common
Country potato, Ubi kemili or kembili (Malay).

Synonyms
Coleus tuberosus Benth.

Notes
A small herbaceous annual with small dark brown tubers clustered around the base of the main stem. Widely cultivated in Southeast Asia but rare in the Pacific. More suited to dry sites. N.B: H.C. Sampson (Kew Bulletin, add. ser. XII, 1936, p 54) distinguishes this species from the Hausa or Madagascar potato *C. dysentericus,* Baker (syn. *C. rotundifolius* (Poir.) Cheval and Perrot) cultivated in East Africa.
C. forksii Briq. (syn. *C. barbatus* Benth.) is common on dry hill sites in northern India and harvested for its fleshy roots which are pickled. This is said by some to be the ancestral species.

References
Sastri, B.N. (1950), ed. The wealth of India. Raw Materials, Vol. 2. Dehli. pp 308–309.

Sheik, Daud, M.S. bin (1947). Ubi kemili (*Coleus tuberosus*). Malayan Agricultural Journal 30, 130–132.

Vide also: N.E. Brown, (1894). Tuberous Labiatae. Kew Bulletin, 1894, pp 10–14.

Colocasia esculenta (L.) Schott.

Araceae

Common
Taro (general in Pacific), Dalo (Fiji), Keladi (Malay), Gabi (Philippines), Kocha (India, Sanskrit), Satoimo (Japan).

Synonyms
Colocasia antiquorum Schott., *Caladium esculentum* Vent.

Notes
A basic starchy root crop of the wet tropics and sub–tropics. From the Pacific Islands to China and Japan. At one time cultivated in North Auckland. A collection made by Doug Yen in the 1950's was no longer maintained at Otara by 1968. But a later collection made by Peter Mathews is still held at the Australian National University, Canberra (D. Yen, (1986), pers. comm.). The plants occasionally cultivated in northern New Zealand today are probably recent introductions. There are two main groups of cultivars:
a) plants producing a large main corm with few cormels (4 to 8), *C.e. var. esculenta,* a diploid $2n = 2x = 28$.
b) plants with a relatively small main corm, often markedly acrid due to oxalic acid crystals, and 15 to 20 cormels *C.e. var. antiquorum* $2n = 3x = 42$.

References

Wang, Jow–Kai. (1983), ed. Taro – a review of *Colocasia esculenta* (L.) Schott and its potentials. University of Hawaii Press, Honolulu. 400 pp.

Lambert, M. (1982). Taro cultivation in the South Pacific. South Pacific Commission, Noumea. Handbook No. 22.

See also Best, E. (1976). Maori Agriculture. 2nd edn. Government Printer, Wellington. pp 233–243.

Best, E. (1922). The taro, *Colocasia antiquorum* . N.Z. Journal of Science and Technology 5(4), 204 – 205.

Cooper, R.C. (1969). Flowering of taro (*Colocasia esculenta* (L.) Schott) in N.Z. Records of Auckland Museum 6, 403–406.

Cordia subcordata Lam.

Ehretiaceae

Common

Banalo (Philippines), Tou (Tahiti), Taunave (Samoa), Kou (Hawaii), Nawanawa (Fiji).

Notes

A small tree widespread in coastal areas of the Pacific from New Guinea to Hawaii. The fruit is occasionally used as an edible nut. But they are small (c.1.5 cm. diam.) and the oil in the endosperm is difficult to extract. The timber is soft, durable and valued for carving in many Pacific Islands.
Cordia sebestena L., the geiger tree and source of the West Indian aloe wood is among 5 species of this genus introduced to Hawaii and possibly elsewhere in the Pacific.

References

Neal, Marie C. (1965). In gardens of Hawaii. Bernice P. Bishop Museum, Special Publication 50, Honolulu. pp 714 – 715.

Petard, P. (1986). Plantes utiles de Polynesie. Editions Haere Po No Tahiti. pp 248 – 249.

Cordyceps robertsii Hook.

Hypocreaceae

Common

Vegetable caterpillar.

Notes

A widespread entomogenous fungus infecting larvae of the noctuid moths, *Porina* spp. in New Zealand. A related species *C. sinclairii* similarly infects cicada larvae. Insect larvae infected with *Cordyceps* are a valued drug in Chinese medicine. (Vide; Pharmacopoeia of the Peoples Republic of China, 11th edition 1977).

References

Probably first recorded by the Rev.R.Taylor of Waimate in 1842, (Tasmanian Journal of Science 1, 307–308). There are many records in the Transactions of the N.Z. Institute, including: W.M. Maskell (1894), 27, 665–668; W.L. Buller (1895), 27, 155–156; C. Fitton (1896), 29, 608–609; H. Hill (1901), 34, 396–401 with plate; G.V. Hudson (1906), 39, 195–196.

Salmon, J.T. (1951). Vegetable caterpillars. Tuatara 4 (1), 1 – 3.

McEwen, F.L. (1963). *Cordyceps* infections. In: E.A.Steinhouse (ed.) Insect pathology. Volume 2. Academic Press, New York. pp 273–290.

Lloyd, C.G. (1915). Synopsis of the *Cordyceps* of Australasia. 11p pamphlet. Cincinnati, Ohio.

See also: Willis J.H. (1959). Australian species of the fungal genus *Cordyceps* (Fr.) Link. with critical notes on collections in Australian herbaria. Muelleria 1, 67–69.

Cordyline fruticosa (L.) Chev.

Agavaceae

Common

Ti (Polynesia), Dracaena palm, Masawe (Fiji).

Synonyms

Cordyline terminalis (L.) Kunth, *Taetsia fruticosa* (L.)Merr., *Dracaena terminalis* L.

Notes

A small species of *Cordyline* with broad lanceolate leaves and a thick starchy tap root for which it was cultivated in Polynesia. There were many cultivars. Petard (1986) lists 16 from Tahiti and the Marquesas. Coloured leafed cultivars have been widely distributed as ornamentals. The species was introduced into New Zealand by the Maori but was rare at the time of European settlement.

References
Cheeseman, T.F. (1901). Notes on the cultivated food plants of the Polynesians, with special reference to Ti Pore (*Cordyline terminalis*). Trans. and Proc. N.Z. Inst. 33, 306–311.

Uhe, G. (1974). Wayside plants of the South Pacific. Stocton House, Auckland. pp 35–36.

Cox, P.A. (1982). Cordyline ovens (umu ti) in Samoa. Economic Botany 36(4), 389–96.

Barrau, J. (1965). Witness of the past. Ethnology 4, 282–294.

For traditional Hawaiian usage of this plant see also: Degener, O. (1945). Plants of Hawaii National Park illustrative of plants and customs of the South Seas. Edwards Brothers, Ann Arbor. pp 95 – 98.

N.B. The use of New Zealand native Cordyline spp. is extensively covered in Elsdon Best, Maori agriculture. Government Printer, Wellington 1976. pp 256–272.

Frankhauser, B.L. (1985). Preparation of high–fructose syrup from New Zealand cabbage tree, *Cordyline australis*. N.Z. Journal of Technology 1, 27–31.

Coriandrum sativum L.

Umbelliferae

Common
Coriander, Dhania (Hindi), Kotomali (Tamil).

Synonyms
Coriandrum testiculatum Lour.

Notes
Two basic types. 1. Small dark fruited cultivars with a high essential oil content are widely grown in Europe and Russia. 2. Cultivars with larger pale straw coloured fruits with low essential oil content are grown in India and the Middle East. Approximately 30% of most curry powders are made up of ground seeds of this latter type. Some crops of coriander have been grown in Canterbury since 1980 both for local consumption and for export, particularly to Thailand.

References
N.B. Most references are of East European/Russian or Indian origin.

Notes on the production of coriander (1962). Tropical Products Institute, London. Report No 66/62. (not sighted).

Purseglove, J.W. et al. (1981). Spices. 2 vols. Longman, London – vide Vol. 2 pp 736–788 (90 references).

Crambe abyssinica Hochst. ex R.E. Fries

Cruciferae

Common
Crambe.

Notes
Interest was shown in this crop in the late 1960's and early 1970's as an oil seed containing 50 to 65% erucic acid in the extracted oil. Trial plots were sown at Lincoln in 1978/9 and possibly elsewhere but there was no follow up.

References
White, G.A. (1966). What we know about growing crambe. Crops and Soils 18(4), 10–12.

White, G.A. and Higgins, J.J. (1966). Culture of crambe, a new industrial oilseed crop. ARS, USDA Production Research Report No. 95. 18 pp.

Crambe maritima L.

Cruciferae

Common
Sea kale, Hamana (Japan).

Notes
Widely used as a vegetable in the 19th century in England and New Zealand; now relatively uncommon. The young shoots are forced in darkness during late winter and early spring.

References
Sinclair, J.T. (1918). Vegetable growing in New Zealand. Whitcombe and Tombs, Auckland. pp 89–91. (Later editions of this work were revised by J.A. McPherson and F.J.E. Jollie but the title remained the same. Sea kale was covered in pp 152–154 in the 6th edition of 1949).

Kennelly, A.G. (1952). The home vegetable garden. N.Z. Department of Agriculture Bulletin 243, pp 148–149.

Figure 18: Seakale shoots.

Figure 19: Saffron crocus.

Crescentia cujete L.

Bignoniaceae

Common
Calabash, Cujeta (Spanish–Philippines), Guiro (Caribbean), Ticaro (Central America).

Synonyms
Crescentia cuneiflora Gardn.

Notes
A tree from tropical America widely cultivated as an ornamental throughout the tropics and for its gourd like fruit which are used medicinally in South and Central America and France. The dried shell of the fruit is traditionally used to make 'maracas', a musical instrument. The calabash is rare in the South Pacific except in New Caledonia.

References
Hawkes, A.D. (1967). Consider the calabash. American Horticultural Magazine 46(2), 90–91.

Heiser, C.B. (1979). The gourd book. University of Oklahoma Press, Norman. pp 15–29.

Crocus sativus L.

Iridaceae

Common
Saffron crocus.

Synonyms
Crocus officinalis Houck.

Notes
Saffron is derived from the dried orange stigmas of this sterile crocus. It is used as a food colourant, medicine and general dye. As a colourant it is now largely replaced by turmeric (*Curcuma domestica*) because of the high price. Saffron is cultivated in Spain, Sicilly and Kashmir in India. This species has been tried many times in New Zealand gardens but seldom flowers here.

References
Palmer, J. (1983). Saffron. Annual Journal of Royal New Zealand Institute of Horticulture 11, 93–97.

International Trade Centre UNCTAD/GATT. Spices – a survey of world markets. Geneva 1982. Vol. 1, pp. 71–72.

Basker, D. and Negbi, M. (1983). Uses of saffron. Economic Botany 37(2), 228–236.

Cuminum cyminum L.

Umbelliferae

Common
Cumin.

Notes
Widely cultivated in south eastern Europe, the Middle East, Egypt and India. A slender much–branched annual requiring a relatively long growing season with the absence of extremes of drought, heat or cold. Cumin seeds are a constituent of curry powders and widely used as flavouring in the food and beverage industry. N.B. 'Black cumin' is the smaller sweeter seed of *Bunium persicum* B. Fedtsch., native to the Middle East Region, or *Nigella sativa* L.(Ranunculaceae), cultivated in Egypt, the Middle East and India.

References
The wealth of India – Raw Materials Vol. 2. (1950). CSIR, Delhi. pp 396–398.

Rosengarten, F. (1969). The book of spices. Livingston Publishing Co., Wynnewood. pp 224–231.

Cuphea wrightii Gray

Lythraceae

Common
Cuphea.

Notes
A short annual species from warm temperate areas of Mexico and southern U.S.A. This seems to be the most promising of a number of *Cuphea* spp. which have seeds high in short chain fatty acids suited to industrial processing.

References
Thomson, A.E. (1985). New native crops for the arid southwest. Economic Botany 39(4), 436–453.

Anon (1985). Cuphea. Diverse fatty acid composition may yield oleochemical feedstock. JAOCS 62(1), 6–12.

Curcuma domestica Lour.

Zingiberaceae

Common
Turmeric, Ago (Samoa), Ango (Tonga), Avea (Fiji).

Synonyms
Curcuma longa Koenig

Notes
Turmeric is a sterile triploid, cultivated as a spice but more particularly as a deep yellow food colourant. India is the major producing country. The rhizomes are usually boiled to reduce drying time and to produce more uniform colouring before peeling, drying and polishing. At one time cow dung was employed in the boiling process to 'improve' the colour. Turmeric is widespread in the islands of the South Pacific but only of importance amongst the Indian population of Fiji and in Samoa where the starch remaining after dye extraction is cooked up with coconut cream. A dish called 'lenga'.

References
Purseglove, J.W. et al. (1981). Spices. Vol. 2. Longman, London. pp 532–580.

Prasad, J. and Singh, A. (1980). Processing of turmeric. Agricultural Journal, Fiji 42(2), 23–25.

N.B. See also: Sopher, D.E. (1964). Indigenous uses of turmeric (*Curcuma domestica*) in Asia and Oceania. Anthropos 59, 93–127.

Curcuma zedoaria (Berg.) Roscoe

Zingiberaceae

Common
Zedoary, Shoti, Indian arrowroot*, Gajutsu (Japan).

Notes
A tropical herb, wild and cultivated throughout Southeast Asia from Sri Lanka to southern China. The rhizomes are a source of starch in India, elsewhere used as minor flavouring or medicine. A large number of related Curcuma species are utilised for similar purposes in Southeast Asia. Zedoary is included in the Japanese Pharmacopoeia (1982).

References
Kay, D.E. (1973). TPI Crop and Product Digest No. 2. Root Crops. Tropical Products Institute, London. pp 135–138.

Prana, M.S. and Danimihargja, S. (1981). Root and tuber crops. IBPGR Secretariat, Rome. pp 48–49.

*N.B. for arrowroot see *Maranta arundinacea* .

Cyamopsis tetragonaloba (L.) Taub

Leguminosae

Common
Guar, Cluster bean, Gowar (Hindi). NB: 'guar' is the Punjabi name for this plant.

Synonyms
Cyamopsis psoraleoides DC.

Notes
Guar is an erect annual legume, up to 2 m in height. In India and Pakistan it is grown for both grain and forage. The young immature pods are edible but they soon become leathery. Elsewhere in arid sub–tropical areas guar is grown as a source of an industrial gum, galactomannan, which occurs in the endosperm of the seed. This is used in the food industry as a thickener, in pharmacy, textiles and in some paints. One cultivar, 'Brooks' tried at Otara in 1974/5 was a failure, possibly due to the relatively low ambient temperature.

References
Jackson, K.J. and Doughton, J.A. (1982). Guar: a potential industrial crop for the dry tropics of Australia. Journal of the Australian Institute of Agricultural Science 481, 17–32.

Whistler, R.L. and Hymowitz, T. (1979). Guar, agronomy, production, industrial uses and nutrition. Purdue University Press, West Lafayette.

Cycas circinalis L.

Cycadaceae

Common
Sago tree, Nan yo sotetsu (Japan), Varoguna (Hindi).

Synonyms
Cycas rumphii Miq. var *seemanii* (A. Braum.) Kanehira, the logo–logo of Fiji may be the same species.

Notes
Cycads are dioecious. Both the large fruit and the starch in the stems are edible but need processing to remove toxins. *C. circinalis* is the principle species used in southern Asia and western Melanesia. *C. revoluta* Thunb., the sotetsu of Japan, has a more northerly range extending to southern Kyushu. *C. media* R. Br., the Australian tree zamia, is rather more toxic but has been utilised by the aborigines in North Queensland.

References
Thieret, J.W. (1958). Economic botany of cycads. Economic Botany 12, 3–33

Whiting, M.G. (1963). Toxicity of cycads. Economic Botany 17, 271–302.

Cyclanthera pedata (L.) Schrader

Cucurbitaceae

Common
Korila, Achoccha (Peru)

Notes
A little known gourd from Central and South America. The fruit are oval, flattened with a sharp pointed tip, used as a vegetable in Mexico and Peru. Herklots (1972) found this species cultivated in Nepal and Taiwan and a specimen was discovered in an Auckland garden in 1979, so it may have penetrated the Pacific.

References
Herklots, G.A.C. (1972). Vegetables in Southeast Asia. Allen and Unwin, London. pp 318–321.

Bristol, M.L. (1966). Native Andean food plants. Newsletter of the Hawaiian Botanical Society 5(3), 19–20.

N.B. References to this species are exceedingly rare and do not occur in many places they might be expected e.g. J.N. Rose (1899). Notes on useful plants in Mexico, or any of the standard works on economic Cucurbitaceae consulted.

A description of the species together with a plate is found in: Pitter H. (1910). New or noteworthy plants from Colombia and Central America. Contributions from the U.S. National Herbarium 13(4), 120–123.

Cymbopogon schoenanthus (L.) Spreng.

Gramineae

Common
Lemon grass, Serai (Malay), Bika (PNG), Harva chella (Hindi), Coboi (Fiji), Moegalo (Samoa), Tanglad (Philippines).

Synonyms
Andropogon cerifera Hook, *A. citratus* DC., *A. citriodorum* Desf., *A. martinii* Wats., *A. roxburghii* Nees, *A. schoenanthus* L., *Cymbopogon citratus* (DC.) Stapf, *Schoenanthum amboinicum* Rumph.

Notes
A perennial tufted grass with erect lemon scented leaves up to 1 m long. Of unknown origin but widely distributed in all warm climatic areas. Occasionally cultivated in N.Z., usually as a pot plant. The leaves and stem bases are used as food flavouring.
N.B. Other Cymbopogon species cultivated for their essential oils include:
Cymbopogon nardus (L.) Rendle (syn. *C. winterianus* Jow.), Citronella grass – a source of citronella oil.
C. martinii Stapf. var. motia Burkill, Rosha grass – for palmarosa oil a source of geraniol.
C. flexuosus (Nees ex Steud.) Wats. Malabar grass.

References
Atal and Kapur, (1982), eds. Cultivation and utilisation of aromatic plants. CSIR, Jammu Tawi. Especially pp 302–307, 318–324.

Morton, Julia (1976). Herbs and Spices. Golden Press, New York. p 11.

Cyperus esculentus L.

Cyperaceae

Common
Chufa, Tiger nut, Earth nut.

Synonyms
Cyperus tenorii Presl.

Notes
Probably Mediterranean in origin. Grown commercially in Spain and the Middle East for the edible tubers. Has been recorded in New Zealand (from Hamilton) but should not be confused with the weedy *C. rotundus* L. which has similar but smaller tubers at the ends of underground stolons. This latter species is used in Chinese medicine.

References
Kay, D.E. (1973). TPI Crop and Products Digest 2. Root crops. Tropical Products Institute, London. pp 50–56.

Coyl, L.P. (1982). World encyclopedia of food. Facts on File, New York. p 169.

Cyrtosperma chamissonis (Schott) Merr.

Araceae

Common
Swamp taro, Via kana (Fiji), Puraka (Cook Is), Maota (Tahiti), Gallan (Philippines)

Synonyms
Cyrtosperma edule Schott, *C. merkusii* (Hasskarl) Schott.

Notes
A large herbaceous perennial (to 2.5 m), leaf blades may be 1.5 m long, with a petiole diameter of 10 cm. Leaf shape and colour variable. Capable of growing in brackish water in coastal bogs and on coral islands where cultivation of Colocasia is difficult. Can be harvested at any time when plants are from 1 to 10 yrs old.

References
Thomson, S. (1982). *Cyrtosperma chamissonis:* ecology distribution and economic importance in the South Pacific. J. Agric. Trad. Bot. Appl. 29, 185–203.

O'Hair, S.K. and Asokan, M.P. (1986). Edible aroids: botany and horticulture. Horticultural Reviews 8, 43–99. Avi Publ., Westport, Conn.

Datura metel L.

Solanaceae

Common
Datura, Black datura, Devil's trumpet.

Synonyms
Datura fastuosa Clarke, *D. alba* Nees.

Notes
A species of medicinal value as a source of atropine. A double purple flowered cultivar is grown as an ornamental plant on some Pacific Islands. The species is occasionally used as a virus indicator plant on N.Z. research stations.

References
Sharman, P.J. (1972). Other crops of potential importance. In Australian Specialist Conference on plants of potential economic importance. Canberra. Sect. 6, pp 3–4.

Sabti S.N. and Kaul, B.L. (1982). Cultivation of *Datura innoxia* and *D. metel* in India. In Atal C.K. and Kapur, B.M. (eds). Cultivation and utilisation of medicinal plants. CSIR, Jammu–Tawi. pp 259–261.

International Trade Centre (1982). Markets for selected medicinal plants. UNCTAD/GATT, Geneva. pp 103–105.

Datura stramonium L.

Solanaceae

Common
Jimson weed, Thorn apple.

Synonyms
Datura stramonium var. tatula L.

Notes
Originally from America but now an ubiquitous weed of warm temperate areas. Occurring spasmodically on both main islands of N.Z. Used medicinally as an antispasmodic in asthma and included in the British Herbal Pharmocopaeia.

References
Avery, A.G., Satina, S. and Rietsema, J. (1959). Blakeslee: the genus *Datura.* Ronald Press, New York. 289 pp.

Connor, H.E. (1977). The poisonous plants of New Zealand. 2nd edn., Government Printer, Wellington. pp 159–161.

Derris elliptica (Roxb.) Benth.

Leguminosae

Common
Derris, Tuba (Malay), Duva (Fiji), Tubli (Philippines), Ora (Cook Is.).

Synonyms
Cylista piscatoria Blanco, *Galattia terminaliflora* Blanco, *Galedupa elliptica* Roxb., *Millettia piscatoria* Merr., *Millettia splendidissima* Vidal, *Pogamia dubia* Grah., *Pogamia elliptica* Wall., *Pogamia volubilis* Zoll. et Mor.

Notes
A woody climber, native to Southeast Asia and introduced to Fiji. But the plant recorded as introduced to many other Pacific Islands is *D. malaccensis* Prain, vide: W.R. Sykes, Contributions to the Flora of Niue. DSIR Bulletin 200, Government Printer, Wellington 1970. pp 150 and 321. The roots are a source of rotenone, traditionally used as a fish poison, but derris is now mainly used as an insecticide of low mammalian toxicity. N.B. *Derris trifoliata* Lour. is native to many islands in the South Pacific and is the traditional fish poison though it contains less rotenone than *D. elliptica* .

References
Duke, J.A. (1981). Handbook of legumes of world economic importance. Plenum Press, New York. pp 73–77. Duke treats both *D. elliptica* and *D. malaccensis.*

Moreau, R.E. (1944). Derris agronomy: An annotated bibliography. East African Agricultural Journal 10(2), 75–82.

Lever, R.J.A. (1938). Derris in Melanesia. Agricultural Journal, Fiji 9(2), 25.

Digitalis lanata Ehrhart

Scrophulariaceae

Common
Woolly foxglove, Grecian foxglove.

Notes
This is the preferred medicinal foxglove yielding the glycoside digoxin which is about twice as biologically active as digitoxin from the common purple foxglove (*D. purpurea*). Trials at Lincoln, 1979–82 produced yields of up to 2,000 kg dried foliage per hectare. However, no markets were found at that time. *D. lanata* is grown in a few N.Z. gardens as an ornamental.

Figure 20: Woolly foxglove plots, Lincoln 1981.

References

Morton, Julia F. (1977). Major medicinal plants. Charles C. Thomas, Springfield. pp 319–321.

Swan, J.B. (1945). Biological assay of Digitalis. N.Z. J. Sci. Technol. B 26(4), 163–174.

Neczypar, W. et al. (1980). Studies concerning obtaining of a high class Digitalis lanata drug. Acta Horticulturae 96(2), 207–209.

Digitaria cruciata Nees. ex Hook. f. *var. esculenta.*

Gramineae

Common
Khasi millet, Raishan (Hindi?).

Notes
The type species is wild in Northeastern India and Southern China. Cultivars are generally taller and adapted to produce quantities of small grain under adverse agronomic conditions.

References

Singh, H.B. and Anora, R.K. (1972). Raishan (*Digitaria* sp.), a minor millet of the Khasi hills, India. Economic Botany 26, 376–380.

Bor, N.L. (1955). The genus *Digitaria* Heist. in India and Burma. Webbia 11, 301–367.

N.B. *Digitaria sanguinalis* Lour., crab grass, an ubiquitous weed of warm temperate areas, is sometimes cultivated as a minor cereal in Southeastern Europe .

Other species of *Digitaria* are cultivated in Africa. Vide: Porteres, R. (1955). Les cereales mineares de genre *Digitaria* en Afrique et en Europe. J. agric trop. Bot. appl. 2, 330–386, 477–510, 620–675.

Dillenia indica L.

Dilleniaceae

Common
Elephant apple, Indian catmon (Philippines), Simpoh (Malay), Chalta (Hindi).

Synonyms
Dillenia speciosa Thunb.

Notes
A small spreading tree with large oblong leaves and large white flowers. The fruit is edible but more usually utilised in curries or for chutney. The bark and leaves are astringent and used medicinally. It is native to Southeast Asia but introduced to the Philippines and Hawaii. It is rare in the South Pacific.

References

Vaidehi, M.P. and Annapurna, M.L. (1985). Beverage from *Dillenia indica* . Indian Horticulture 30(1), 14.

The wealth of India. Raw materials . CSIR New Delhi, 1952. Vol. 3. pp 64–65.

Dioscorea alata L.

Dioscoreaceae

Common
Yam, Greater yam, White yam, Ubi (Malasian archipelago), Uhi (Hawaii, Tahiti), Ufi (Samoa, Niue), Uvi (Fiji).

Synonyms
D. eburnea Willd., *D. globosa* Roxb.

Notes
A strong growing vine. Stems with membranous wings and opposite, caudate leaves. The tubers are large and variable in shape and colour. A species probably originating in Southeast Asia but widespread in the South Pacific and at one time grown in northern New Zealand (Best, 1976).

References

Martin, F.W. (1976) Tropical yams and their potential Pt 3. *Dioscorea alata.* USDA Handbook No. 495, Washington. 40 pp.

Best, E. (1976). Maori agriculture. 2nd edn. Government Printer, Wellington. pp 228–232.

Coursey, D.G. (1967). Yams. An account of the nature, origins, cultivation and utilisation of the useful members of the Dioscoreaceae. Longmans, London. 230 pp.

N.B. Some 6 species of yams are widely cultivated in the South Pacific. Apart from *D. alata,* these include *D. bulbifera, D. pentaphylla, D. esculenta, D. hispida* and *D. nummularia.* vide: Massal, E. and Barrau, J. (1956). Food plants of the South Sea Islands. South Pacific Commission Technical Paper No. 94. pp 12–15.

Dioscorea opposita Thunb.

Dioscoreaceae

Common
Chinese yam, Naga–imo (Japan).

Synonyms
Dioscorea batatas Decne.

Notes
One of the most hardy species of *Dioscorea* , cultivated as far north as Japan. The long club shaped root is a minor food crop in the Far East and is also considered medicinal. It is included in the Japanese Pharmacopoeia along with the native *D. japonica* Thunb. Introduced for trial purposes into New Zealand in 1986. (J. Palmer, CRD Lincoln).

Figure 21: Chinese yam tubers (c. 60 cm. long). The remains of previous years tuber on left.

References

Herklots, G.A.C. (1972). Vegetables in Southeast Asia. George Allen & Unwin, London. pp. 422–423.

Kawakami, K. (1970). Yam culture in Japan. Proc. 2nd International Symposium on Tropical Root and Tuber Crops Vol. 1 Part 3, 102.

Palmer, J. (1988). The Chinese yam (*Dioscorea opposita* Thunb.). Annual Journal of the Royal N.Z. Institute of Horticulture 15, 36–38.

Dovyalis hebecarpa Warb.

Flacourtiaceae

Common
Ceylon gooseberry, Ketambilla (Singhalese), Aberia (Cuba).

Synonyms
Aberia gardneri Clas., Roumea hebecarpa Gard.

Notes
A small spiny tree native to hill areas of Sri Lanka and introduced to other tropical areas. Cultivated in the Philippines and Hawaii. Scarce in the South Pacific but known from Fiji and northern New Zealand. The brown/purple velvety fruit, 2.0 to 2.5 cms diam. is usually very acid and most suited to processing.
A hybrid between this species and the Ethiopian, *D. abyssinica* (A. Rich.) Warb. has been marketed in Florida and introduced to Queensland.
Dovyalis caffra (Hook. f. and Harv.) Warb. The kei apple of southern Africa is grown in the Philippines but is rare in the South Pacific. It has been recorded in New Zealand. Vide: Trap L.(1986). *Dovyalis caffra* . Growing Today 4(3),39.
N.B. Dovyalis is occasionally spelt Doryalis.

References
Parsons, T.H. (1937). The cultivation of fruits in Ceylon. Department of Agriculture, Bulltin No. 90, Colombo. p 18.

Morton, Julia F. (1987). Fruits of warm climates. Creative Resource Systems Inc., Winterville, N.C. pp 311–313.

Matheson, E.M. (1979). Other crops of potential importance. In, Lovett, J.V. and Lazenby, A. (eds.) Australian field crops. Angus and Robertson, London and Sydney, Vol. 2, pp 263–286.

Duboisia spp.

Solanaceae

Common
Corkwood.

Notes
Three species are native to Australia:
Duboisia myoporoides R.Br. has an extensive range in coastal sites from New South Wales to Northern Queensland.
D. leichhardtii F. Muell. is limited to a small area of southeast Queensland.
D. hopwoodii F. Muell. is widespread in arid regions of the interior. It is chewed by aborigines and contains up to 7% nicotine. Vide: Curl, S.M. (1878). On 'pituri' a new vegetable product that deserves further investigation. Transactions of the N.Z. Institute, 11, 411–415.
The first two species are cultivated as a source of the pharmaceutical alkaloids hyoscine and hyosamine. Dried foliage is exported mostly to Japan and West Germany.

References
Loftus–Hills, K. et al. (1954). Variation in the main alkaloids of *Duboisia myoporoides R.* Br. and *D. leichhardtii* F. Muell. Australian Journal of Applied Science 5, 258–275.

Durio zibethinus Murr.

Bombacaceae

Common
Durian.

Synonyms
Durio foetida Thunb.

Notes
A large tree from tropical Southeast Asia. Introduced to Papua New Guinea and Tahiti. Rare elsewhere in the Pacific. Fruits may weight up to 3 kg.

References
Malo, S.F. and Martin, F.W. (1979). Cultivation of neglected tropical fruits with promise, Pt 7, The durian. USDA, New Orleans.

Watson, B.J. (1984). Durian. In P.E. Page (ed.) Tropical tree fruits for Australia. Queensland Department of Primary Industries, Brisbane. pp 45–50.

Ecballium elaterium (L.) A. Rich.

Cucurbitaceae

Common
Squirting cucumber, Elaterium (Pharmaceutical).

Synonyms
Momordica elaterium L.

Notes
A scrambling herb from the Mediterranean area with a deep tap root which tends to be perennial on suitable sites. The fruit are violently explosive. The sediment from the juice of the fruit, dried into opaque cakes is known as 'elaterium' and was once used as a strong cathartic and it is occasionally still traded. Malta is a major source. The squirting cucumber is a weed in eastern Australia and has been recorded wild in New Zealand (Hauraki Plains).

References

Everist S.L. (1981). Poisonous plants of Australia. Revised edition, Angus and Robertson, Sydney. p 223.

The Wealth of India – Raw Materials. CSIR, New Delhi 1952. Vol. 3. p. 123.

Echinacea angustifolia (DC.) Heller

Compositae

Common
Cone flower, Black sampson, Echinacea.

Synonyms
Broaneria angustifolia Nectar., *Echinacea pallida var. angustifolia* (DC.) Cronquist.

Notes
A small herbaceous perennial from the prairie states of North America. The flowering stems up to 60 cm high are sparsely or densely covered with rough bristly hairs. The dried root is widely used in herbal medicine but according to Foster (1985) "much of the *E. angustifolia* on the American herb market is suspected to be *E. pallida* Nutt."

References

Foster, S. (1985). Echinaceae exalted. 2nd ed. Ozark Beneficial Plant Project, Brixy, Miss. 40 pp.

Beaubaire, N. and Simon, J.E. (1987). Field establishment of three *Echinacea* species for commercial production. Acta Horticulturae 208, 115–120.

Echinochloa spp.

Gramineae

Common
Echinochloa colona (L.) Link, Awnless barnyard grass (Australia); Sawa millet, Sawet, Swank (Hindi);
E. frumentacea Link, Siberian millet, Sanwa (Hindi).
E. turnerana (Domin.)J.Black, Channel millet.
E. utilis Ohwi and Yabuno, Japanese millet.

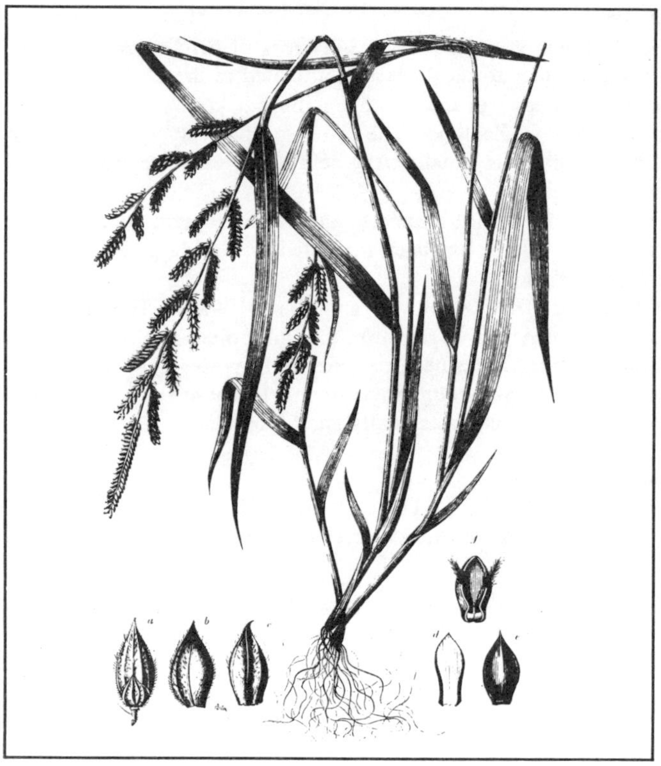

Figure 22: Sawa millet.

Notes
All the above species have been recorded from Australia. Most have been grown at some time in N.Z. *E. turnerana* is native to south–western Queensland. *E. frumentacea* and *E. colona* are cultivated in India.
Several other closely related species of *Echinochloa* are grown as small grain crops.

References

De Wet J.M.J. et al. (1983). Domestication of sawa millet (*Echinochloa colona).* Economic Botany 37(3), 283–291.

National Academy of Sciences (1975). Underexploited tropical plants with promising economic value. NAS, Washington. pp 9–13 (on *E. turnerana.*)

Tabuno, T. (1987). Japanese barnyard millet (*Echinochloa utilis*). Economic Botany 41(4), 484–493.

N.B. 'Common' millet is *Panicum miliaceum.* For other species of 'millet' see: *Digitaria, Eleusine* and *Paspalum* .

Edgeworthia papyrifera Sieb. and Zucc.

Thymelaeaceae

Common
Paper bush, Mitsumata (Japan), Chi hsiang (China – literally 'knot scent').

Synonyms
Edgeworthia chrysantha Lindl., *E. tomentosa* Nakai, *Daphne papyrifera* Sieb.

Notes
A shrub with pendulous clusters of sweet–smelling yellow flowers appearing in early spring. The stems are markedly pliable, hence the Chinese name. Native to China but cultivated in southern Japan as a source of soft tough handcrafted paper prepared from the bark. It is used for legal documents, diplomas and bonds. Occasionally grown in New Zealand gardens as an ornamental but could be exploited for craftwork.

References
Fairchild, D.G. (1903). Three new plant introductions from Japan – Mitsumata, a Japanese paper plant. USDA, Bureau of Plant Industry, Washington, Bulletin No. 42. pp 9–17.

Bean, W.J. (1973). Trees and shrubs hardy in the British Isles. 8th edition. John Murray, London Vol. 2. pp 74–75.

Eleocharis dulcis (Burm.f.) Trin. ex Henschel

Cyperaceae

Common
Chinese water chestnut*, Matai, Kurogawai (Japan), Kuta (Tonga), Kohekohe (Hawaii).

Synonyms
Eleocharis tuberosus Schuet., *Scirpus tuberosus* Roxb.

Notes
An aquatic sedge with tuberous roots. Cultivated in China and to a limited extent in the U.S.A., Hawaii and Tahiti.

References
Hodge, W.H. (1956). Chinese water chestnut or matai. Econ. Bot. 10(1), 49–65.

Williams, A.K. and Leeper, G.F. (1976). Chinese water chestnut culture. USDA Information Sheet, Horticulture Crops Laboratory, Athens, GA.

Twigg, B.A. (1957). Cultural studies with matai (Chinese water chestnut). American Society for Horticultural Science 70, 266–272.

*N.B. There is a confusion of names here as *Trapa bicornis* is also cultivated in China and is called 'water chestnut'.

Elettaria cardamomum (Roxb.) Maton

Zingiberaceae

Common
Cardamom, Elachi (Hindi), Katamone (Samoa).

Synonyms
Alpinia cardamomum Roxb., *Amomum cardamomum* Lour., *A. racemosum* Lam., *A. repens* Sonner.

Notes
A large (to 5 m) herbaceous plant, native to southern India and Sri Lanka. Cultivated for the seed pods borne on long trailing basal sprays. About 70% of world production comes from India. Other producers include Guatemala, Sri Lanka and Papua New Guinea. Cardamon has been introduced to several islands in the South Pacific but is rare. Commercial trials in Fiji before World War 2 failed.
The two major markets for cardamons are Arab countries of the Middle East where it is a major flavouring agent in coffee. And Scandanavian countries where it is a traditional flavouring of baked goods. Cardamons are a high priced spice and there are numerous substitutes – see below.
N.B. Other 'cardamoms' of commerce are usually designated 'false cardamom' to distinguish them from the 'true cardamom' above, include:
Amomum aromaticum Roxb. Bengal cardamom, Morang elaichi (Hindi) is cultivated from the Himalayas to Vietnam.
Amomum subulatum Roxb. Greater Indian or Nepal cardamom, Bara elachi (Hindi) is grown in the hill areas of Sikkim and Nepal, and quite large quantities are exported via Singapore.
Amomum krervanh Pierre, from Thailand and Cambodia.
Amomum kepulaga Sprague and Burkill, is the round or cluster cardamom from Java (Syn. A. cardamomum Willd.).
Amomum globosum Lour. The large round cardamom from southern China and used in Chinese medicine.
Amomum xanthioides Wallich. Tavoy or bastard cardamom from Cambodia and Thailand, much used in Thai cooking and also in Chinese and Japanese medicine.
Aframomum mala K. Schum. Wild or Madagascar cardamom from East Africa.
Aframomum korarima from Ethiopia is also marketed as korarima cardamom.

References

Purseglove, J.V. et al (1981). Spices. Longman, London. Vol. 2. pp 581–643.

Sarath Kumara, S.J. et al (1985). Some studies on the effect of maturity and storage on the chlorophyll content and essential oils of the cardamom fruit. Journal of Science, Food and Agriculture 36(6), 491–498.

Eleusine coracana (L.) Gaertn.

Gramineae

Common
African millet, Indian millet, Finger millet, Ragi, Mandua (Hindi–Fiji), Koracan.

Synonyms
Eleusine coracan (L.) Aschers. and Graebner

Notes
A small grain crop of hill areas of northeastern India where it tends to be replaced on poorer soils by kodo millet (*Paspalum scrobiculatum*). The grain is often fermented to make an alcoholic beverage in the Himalayan area. Introduced and occasionally cultivated in Fiji and in eastern Australia. Also grown in East Africa.

References

Purseglove, J.W. (1972). Tropical Crops – Monocotyledons. Longman, London. Vol. 1. pp 147–156.

Hila, K.W. and de Wet, J.M.S. (1976). Domestication of *Eleusine coracana.* Economic Botany 30, 199–208.

Eleutherococcus senticosus Maxim.

Araliaceae

Common
Eleuthero, Siberian ginseng, Devil's shrub, Netronnik (Russian).

Synonyms
Acanthopanax senticosus Harms.

Notes
A spiny sub–shrub from north China through into Siberia between 35 and 53 deg N latitude. Harvested from wild stands in the USSR. Occasionally cultivated. Seed is available from Canada and Tasmania and has recently been introduced to N.Z. The medicinal use of this adaptogen in stressful situations is similar to that of ginseng.

References

Farnsworth, N.R. et al (1985). Siberian ginseng (*Eleutherococcus senticosus):* Current status as an adaptogen. In H. Wagner, Hiroshi Hikino, and N.R. Farnsworth (eds.) Economic and Medicinal Plant Research. Academic Press Inc., Orlando, Vol. 1. pp 155–215.

Fulder, S. (1980). The drug that builds the Russians. New Scientist, 21 August 1980, pp 576–577.

Ensete ventricosa (Welw.) Cheesman

Musaceae

Common
Abyssinian banana

Synonyms
Musa ensete (Gmel.) Horen.

Notes
One of the hardiest of the bananas. A staple food in highland areas (1,600 to 3,000 m) in Southwest Ethiopia where the pseudostems are cut and cooked, sometimes after several days fermentation. The fruit is dry and inedible. Sometimes cultivated as an ornamental in New Zealand.

References

Purseglove, J.W. (1972). Tropical crops – Monocotyledons. Longman, London. Vol. 2. pp 343–344.

Bezuneh, T. and Feleke, A. (1966). The production and utilisation of the genus *Ensete* in Ethiopia. Economic Botany 20, 65–70.

Ephedra sinica Stapf

Ephredraceae

Common
Ephedra, Ma–huang.

Notes
Low branched sub–shrubs. Leaves are reduced to scales so the stem performs the work of assimilation. Of very ancient use in Chinese medicine and the original source of 'ephedrine' now produced synthetically. Grown in New Zealand during World War 2 in the Hutt Valley area but has not persisted.
Three other species are included in the British Herbal Pharmacopoeia (1983) as alternative sources of ephedrine. *E. equisetina* Bunge from China. *E. gerardiana* Wall. (syn *E. distachya* L. and *E. vulgaris* Rich. p.p.) *E. major* Host. (syn *E. nebrodensis* Tines.). The last two species are from hilly areas of northern India.

References
Hu Shiu–ying (1969). Ephedra in new Chinese materia medica. Economic Botany 23, 346–351.

UNESCO (1960). Medicinal plants of the arid zone. Paris. pp 32–34.

Eruca sativa Mill.

Cruciferae

Common
Roquette, Garden rocket*, Taramira (Hindi).

Synonyms
Brassica eruca L.

Notes
An annual, native to the Mediterranean region where it is a salad vegetable. The same species is an oil seed in India. Widespread in N.Z. gardens but under utilised as a winter salad. References to this excellent salad are extremely rare in the English language.

References
Organ, J. (1960). Rare vegetables for garden and table. Faber & Faber, London. pp 150–151.

Gorini, F (1979). Ortaggia da foglia 2.26, Rucula o ruchetta o rughetta. Informatore di Ortoflorofrutticoltura 20 (11), 5–6.

The Wealth of India – Raw Materials, Vol 3. CSIR, New Delhi 1952, pp 190–192. (as an oil seed).

*N.B. Rocket or dame's violet is *Hesperis matronalis*.

Eriobotrya japonica (Thunb.) Lindl.

Rosaceae

Common
Loquat, Japanese medlar. Lokat (Hindi), Loketi (Tonga).

Notes
A small evergreen fruit tree from central China. Introduced into Japan at an early period and cultivated in Europe since the 18th century. Essentially a sub–tropical tree, but hardy in N.Z. as far south as Christchurch.

References
Anon, (1979). Annotated bibliographies: Loquat (*Eriobotrya japonica*),1967–1978. Plant Ser. ISSN 0141–5921, C.A.B., Farnham Royal.

Blumenfeld, A. (1980). Fruit growth of loquat. J. Amer. Soc. hort. Sci. 105 (5), 747–750.

Morton, Julia F. (1987). Fruits of warm climates. Creative Resource Systems Inc., Winterville, N.C. pp 103–108.

Erythroxylon coca Lam.

Erythroxylaceae

Common
Coca, Cocaine bush, Bolivian or Huanuco leaf.

Synonyms
E. bolivianum Burck. = *E. coca* Lam. var. *bolivianum* Burck.

Notes
A shrub from the northern Andean regions of South America long used as a stimulant by the native people, now widely abused as a narcotic. It is cultivated in Java and India for medicinal purposes, and was introduced to some Pacific Islands. The Peruvian species *E. truxillense* Rusby, known in the trade as truxillo leaf, is also accepted in the British Pharmaceutical Codex (1963) as a source of cocaine. The related species *E. novo–caledonicum* Schultz occurs in New Caledonia, but its utility is unknown.

References

Martin, R.T. (1970). The role of coca in the history, religion and medicine of South American Indians. Economic Botany 24, 422–438.

Morton, Julia. (1977). Major medicinal plants. Charles C Thomas, Springfield, Ill. pp 177–183.

Eugenia uniflora L.

Myrtaceae

Common
Surinam cherry, Cayenne cherry, Chermai belanda (Malay), Pitanga (Hindi).

Synonyms
Eugenia michelii Lam., *Stenocalyx brasiliensis* Berg.

Notes
A small South American tree with small scarlet edible fruit. Widely planted in the tropics but mostly grown in home gardens. Relatively rare in the South Pacific and tropical Australia. N.B. Not to be confused with the West Indian cherry (*Malpighia punicifolia*).

References

Shamel, A.D. and Popenoe, W. (1916). The pitanga. Journal of Heredity 7, 179–185.

Molesworth–Allen, Betty (1967). Malayan fruits. Donald Moore Press, Singapore. pp 112–114. Another edn. Longman, London 1975, unsighted.

Creasy, R. (1986). The gardener's handbook of edible plants. Sierra Club Books, San Francisco. pp 274–276.

Euphorbia hirta L.

Euphorbiaceae

Common
Asthma plant, Pill–bearing spurge, Dudhi (Hindi), Eaea (Tahiti), Botabotanis (Philippines), Deniose or Ovoku (Fiji), La'au fai moti (Samoa).

Synonyms
Euphorbia capitata Wall., *E. genella* Lag., *E. globulifera* Kunth., *E. nodiflora* Steud., *E. obliterata* Jacq., *E. ophtalmica* Pers., *E. pilulifera* L., *E. procumbens* DC., *E. verticillata* Vell.

Notes
An annual herb from tropical America. Widespread in the South Pacific as far south as the Kermadec Islands, but only once recorded on mainland New Zealand. Very widely used in herbal medicine and included in the British herbal, Japanese and Chinese Pharmacopoeias.

References

The Wealth of India – Raw Materials, Vol. 3. CSIR, New Delhi 1952. p 225.

Quisumbing, E. (1978). Medicinal plants of the Philippines. Katha Publishing Co., Quezon City. pp 504–507.

Euphorbia lathyris L.

Euphorbiaceae

Common
Caper spurge, Garden spurge, Mole plant, Gopher plant.

Notes
A European biennial herb. Widespread and occasionally cultivated as an ornamental in New Zealand. In the U.S.A. where it has also been introduced it is believed to discourage burrowing mammals giving rise to the last two common names. Recently this spurge has been investigated as an energy crop.

References

Kingsloves, B.E. (1982). *Euphorbia lathyris* reconsidered : its potential as an energy crop for arid lands. Biomass 2, 281.

Connor, H.E. (1977). The poisonous plants in New Zealand. Second revised edition. Government Printer, Wellington. pp 74–76.

See also: Maxwell, B.D. et al. (1985). Energy potential of leafy spurge *(Euphorbia esula).* Economic Botany 39(2), 150–156.

Fagopyrum esculentum Moench

Polygonaceae

Common
Buckwheat, Brank.

Synonyms
F. sagittatum Gilib., *F. vulgare* T.Nees, *Polygonum fagopyrum* L.

Notes
An occasional cereal crop in N.Z. with plants escaping to the roadside verges of both main islands. Buckwheat has a relatively small root system and is adapted to growth under cool moist conditions. The leaves and flowers are a source of 'rutin', a glycoside used medicinally to strengthen weak blood vessels. Japan imports significant quantities.

References
De Jong, H. (1972). Buckwheat – a review. Field Crop Abstracts 25(3), 389–396.

Matheson, E.M. (1979). Other crops of potential importance. In J.V. Lovett and A. Lazenby (eds.). Australian field crops. Angus and Robertson, London. Vol. 2. pp 263–286.

Other species: *Fagopyrum tataricum* Gaertn. Tartarian buckwheat is an even more hardy species cultivated in Nepal and East Asia.

F. cymosum Meissn. (syn. *Polygonum multiflorum* Thunb.), the perennial wild buckwheat of China is used medicinally and is included in the Chinese Pharmacopaeia (1977).

Ferula asa–foetida L.

Umbelliferae

Common
Asa–foetida, Giant fennel.

Synonyms
Scarodasma foetidum Bunge.

Notes
A short–lived perennial herb from arid upland sites of the Iranian–Afganistan region. The top of the root is cut off when the plant reaches the early flowering stage and the exuded gum scraped off to become the asa–foetida of commerce. This is widely used in Arab countries both as a flavouring and in medicine. In Europe it is an essential ingredient of such products as 'Worcester sauce'. Other closely related species such as *Ferula alliacea* Boiss., the 'hing' and *F. foetida* Regel, 'hingra' are used in India. Seed of *F. asa–foetida* was introduced to New Zealand in 1985 and trials are underway at Tara Hills Research Station.

References
UNESCO (1960). Medicinal plants for the arid zone. Paris. pp 37–38.

Chamberlain, D.F. (1977). The identity of *Ferula asa–foetida* . Notes of the Royal Botanic Gardens, Edinburgh 25(2), 229–234.

Ficus carica L.

Moraceae

Common
Fig.

Notes
A fruit tree native to the Mediterranean region where the fruit are normally pollinated by insects (*Blastophaga spp.*). As these insects are not present in N.Z. only those cultivars which set parthenocarpic fruit can be grown. Regrettably the small collection of fig cultivars planted in the DSIR Resarch Orchard at Te Puke were pulled out in 1982.

References
Hiam, N. (1985). Figs – easy to grow, but almost unknown here. Growing Today 2(6), 28–29.

Condit, I.J. and Endesud, J. (1956). The bibliography of the fig. Hilgardia 25, 1–663.

Grassi, G. (1982). The fig : scientific findings from Italy during the last 30 years. Frutticoltura 44 (9/10), 7–10.

Armstrong, W.P. (1987). The fig and its wasp. Pacific Horticulture 48(2), 30–34.

N.B. Amongst other edible species of the genus *Ficus* should be mentioned *Ficus sycomorus* L., the sycomore of the bible. Vide: Harlan, J.R. (1986). Lettuce and sycomore. Sex and romance in ancient Egypt. Economic Botany 40(1), 4–15.

Leaves of several *Ficus spp.* are used as traditional vegetables in Papua New Guinea. Vide: Kurika, L. (1985). Food crop germplasm collection at Keravat. IBPGR Regional Committee for Southeast Asia, Newsletter 9(1), 8–9.

Finschia chloroxantha Diels

Proteaceae

Common
Sauge (PNG), Taroa (New Britain).

Synonyms
Finschia waterhouseana B.L. Burt, *Grevillea densiflora* .

Notes
Finschia is a genus of large trees often buttressed and raised on stilts. From tropical rain forests of the Southwest Pacific but not extending into Australia. The seeds are edible. 'Sauge' is cultivated on village margins from the Caroline Islands southwards through New Guinea, Solomon Islands and Vanuatu.

References
White, C.T. (1949). *Finschia* – a genus of nut trees of the Southwest Pacific. Pacific Science 3(3), 187–194.

Henty, E.E. (1982). Some nut bearing plants in Papua New Guinea. Proceedings of the Second Papua New Guinea Food Crops Conference. Port Moresby. pp 78–85.

See also: Sleumer, H. (1955). Proteaceae. In Flora Malesiana Ser. 1. Vol. 52. pp 159–164.

Other species: *Finschia ferruginiflora* White, 'koparu seeds' are used in New Guinea.

Frangula purshiana (DC.) Cooper

Rhamnaceae

Common
Cascara sagrada, Californian buckthorn, Alder buckthorn.

Synonyms
Rhamnus purshiana DC.

Notes
A small deciduous tree from western North America. Recorded in New Zealand as growing in the Golden Downs Forest area of Nelson. The bark – cascara sagrada, is included in the 19th edition of the US Pharmacopoeia (1975) as a cathartic. Bark obtained from cultivated trees in the United States is considered to be of a lower grade than that derived from feral trees. When harvesting, trees are usually cut to within 15 cm of the ground using a sloping cut, and allowed to re-sprout. Traditionally dried bark is stored for at least one year before use.
In Great Britain the related *Frangula alnus* Mill. (syn *Rhamnus frangula* L.) replaces this species in the Pharmaceutical Codex (1973) and has been official in that country since 1650.
Frangula alnus is native to Europe, Asia and North Africa and is rarely cultivated in New Zealand. The evergreen buckthorn which is a common adventive in many parts of New Zealand is *R. alaternus* L.

References
Morton, J.C. (1977). Major medicinal plants. Charles C. Thomas, Springfield. pp 200–204.

Duke, J.A. (1985). Handbook of medicinal herbs. CRC Press, Boca Raton. p 403.

N.B. Several species of *Rhamnus* are used as dye plants: *R. cathartica* from the Mediterranean region whose berries are purgative. The juice of the fruit when mixed with alum and evaporated forms the paint known as 'sap-green'. *R. infectoria* L., the yellow berries give green and yellow dyes in France. *R. chlorophora* Decne, a Chinese species the bark of which yields Chinese green indigo used in dyeing silk.

Gentiana lutea L.

Gentianaceae

Common
Yellow gentian, Felwort, Bitterwort.

Synonyms
Asterias lutea Borck.

Notes
A slow maturing herb from alpine regions of southern and central Europe. The plant normally forms a rosette for several years before flowering. The roots have been long used as a source of medicinal bitters and 'radix gentianae' is included in most pharmacopoeias for this purpose. Small quantities are also used in flavouring drinks, e.g. Angostura bitters. Cultivation of this species at sea level is reputedly difficult. Problems with seed germination have occurred in plant introductions to New Zealand.
N.B. Other species of gentian are used as alternatives in specific areas, particularly *Gentiana scabra* Bunge, in the Far East. Hybrids of this latter species are cultivated as cut flowers in Canterbury and elsewhere in New Zealand. Vide: John Aldsworth. New Zealand Gardener, November 1986, p4.

References
Barralis, G. et al. (1978). New trends of *Gentiana lutea* cultivation. Acta Horticulturae 73, 303–306.

Franz, Ch. and Fritz, G. (1978). Cultivation aspects of *Gentiana lutea* L. Acta Horticulturae 73, 307–314.

Schultze, J. and Franz, Ch. (1982). Sugars, bitter substances and essential oils of *Gentiana lutea* in dependence of ecotype, plant age and stage of development. Acta Horticulturae 96(1), 311–315.

Gevuina avellana Mol.

Proteaceae

Common
Chilean hazel, Chilean macadamia (N.Z), Avellano, Gevuin.

Synonyms
Quadria heterophylla R. et P., *Q. avellana* (Mol.) C.F.Gaertn.

Notes
A small evergreen tree (to 12 m) from rainforest areas of southern Chile, 35 to 44 degrees South. The racemes of white flowers followed by brick–red drupes make this a useful ornamental plant, but the weak root system renders young plants difficult to transplant. The avellano was introduced to New Zealand as a potential nut crop around 1980. Seedlings did not thrive in a partly sheltered site at Lincoln but others have been recorded as fruiting in Taranaki.

References
Davies D.J.G.and Macfarlane R.P.(1986). Avellano or Chilean macadamia. Growing Today 3(5), 62.

Davies D.J G.(1987). Chilean wild nut – ornamental and edible. Growing Today 4(3), 34–35.

Glehnia littoralis F. Schmidt.

Umbelliferae

Common
Hama bofu (Japan), Silver top, Corkwing, Bei sha shen (Pinyin).

Synonyms
Phellopteris littoralis Benth. et Hook.

Notes
A short–lived perennial species with a wide natural range in coastal areas on both sides of the North Pacific. On the American side the species is rare and endangered. In Japan young shoots blanched in spring are a traditional vegetable. The root is widely used in Chinese medicine and is included in both the Chinese (1977) and Japanese (1982) Pharmacopoeias. Introduced to N.Z. in 1985 and currently in trials at Lincoln.

References
Perry, Lily, M. (1980). Medicinal plants of Southeast Asia. MIT Press, Cambridge. p 416.

Yashiroda, K. (1968). A dictionary of popular Japanese herbs and their uses today. Plants and Gardens 24(2), 44–58.

See also: Takashima, S. et al. (1980). Vegetable crops of Japan in colour. Haikusha Publishing Co., Osaka. p 19. (Text in Japanese).

Glycyrrhiza glabra L.

Leguminosae

Common
Licorice or liquorice, Kanzo (Japan)

Notes
G. glabra is native to southern Europe. It is replaced by *G. glabra var glandulifera* Reg. et Herd., further east and by *G. uralensis* DC. in Chinese medicine. Licorice is a perennial deep rooted herbaceous plant with long rhizomes up to 1.5 cm in diameter and extending more than a metre. The roots and the rhizomes are harvested after a period of three or more years. They are dried with or without peeling and marketed as a source of the licorice of commerce.

References
International Trade Centre (1982). Markets for selected medicinal plants. UNCTAD/GATT, Geneva. pp 133–144.

Palmer, J. (1987). Licorice – lessons from Lincoln trials. Dittany 8, 46–49.

Figure 23: Hama bofu in 15 cm. pot.

Figure 24: *Glycyrrhiza glabra* flowering, December 1988.

Gnetum gnemon L.

Gnetaceae

Common
Tulip (Papua New Guinea – from the English 'two leaf'), Meninjau (Malay), Sikau or sukau (Fiji).

Notes
A slender tree from Assam through Malesia to the Caroline Islands and Fiji. The young leaves are used as a vegetable in Indonesia, Papua New Guinea and occasionally in Fiji. The red fruit is usually sun dried before roasting with or without the peel.

References

Bourke, R.M. (1981). Some little known food crops from Papua New Guinea. Science in New Guinea 8 (3), 164–169.

Ochse, J.J. (1977). Vegetables of the Dutch East Indies. Australian National University Press, Canberra. pp 296–298.

Duke, J.A. (1989). Handbook of nuts. CRC Press, Boca Raton. pp 166–167.

Guizotia abyssinica (L. f.) Cass.

Compositae

Common
Niger, Inga, Ramtil or Rantil seed.

Notes
An edible oil seed crop from East Africa and India. Has been grown experimentally in Australia and small quantities of the shiny black seeds are imported into N.Z. annually as birdseed. The plant is like a small (to 1.5 m) branched sunflower. It is an annual which flowers in short days (?).

References

Jackson, K.J. and Berthelsen, J.E. (1986). Guayule and Niger : little known crops with unusual uses. Queensland Agricultural Journal 112(3), 137–144.

Tripathi, M.D. and Saraswat, B.L. (1984). A study on some important parameters of niger crop. Agricultural Science Digest 4(1), 15–16.

Gynostemma pentaphyllum (Thunb.) Makino

Cucurbitaceae

Common
Ama–cha–zuru* (Japan, literally – sweet tea vine).

Synonyms
Gynostemma pedata Blume, *Vitis pentaphyllum* Thunb.

Notes
A slender scandent perennial, very widespread in occurrence from Korea and Japan south through China to Northern India and Malaysia. The leaves and shoots have a sweet flavour when dried and are used in herbal teas in Japan and Okinawa. This species was introduced into New Zealand in the early 1980's as a private commercial project for re–export to Japan but is of no known medicinal value.

References
Tanaka, T. (1976). Tanaka's cyclopedia of edible plants of the world. Keigaku Publishing Co., Tokyo. p 349.

All other agronomic references sighted are in Japanese. For full description of the plant see:

Walker, E.H. (1976). Flora of Okinawa. Smithsonian, Washington. p 998.

*'Ama–cha' strictly speaking refers to *Hydrangea macrophylla* var. *thunbergii* which is also used as a sweet herbal tea and is official in the Japanese Pharmacopaeia (1982).

Hedychium coronarium Koenig

Zingiberaceae

Common
Ginger lily, White ginger, Kamia (Philippines), Cevuga (Fiji), Tevunga (Tonga), Teuila (Samoa).

Synonyms
Hedychium flavum Roxb., *H. lingulatum* Hassk.

Notes
Native to northern India and southern China, but now pan–tropical. The flowers are sweet scented and the fragrance (Longosa oil) is occasionally used in the Philippines, Madagascar, Tonga and India. Elsewhere the species has been tried for paper making in Brazil and is occasionally used in native medicine in Southwest Asia. The closely related wild ginger, *H. gardnerianum* Ker-Gawl. is common on the west coast areas of both main islands of New Zealand.

References
Anon (1912). New sources of paper. Kew Bulletin No. 9, 373–378.

Turrill, W.B. (1914). *Hedychium coronarium* and related forms. Kew Bulletin No. 68, 368–372.

See also; Winters, H.F. and Corbett, E G. (1965). A cold hardy ginger lily. American Hoticultural Magazine 44(3), 129–131. *(H. spicatum)* .

Helianthus tuberosus L.

Compositae

Common
Jerusalem artichoke; Kiku–imo (Japan).

Notes
Widespread in gardens but seldom commercially cultivated in N.Z. Some interest shown during the energy crisis of the 1970's and a hybrid line imported from Russia (vide J. Palmer, CRD, Lincoln) other accessions imported privately from U.S.A.

References
Boswell, V.R. (1959). Growing Jerusalem artichoke. USDA Leaflet No. 11, Washington.

Heiser, C.B. (1976). The sunflower. University of Oklahoma Press, Norman. 198 pp.

Vide also: The Press 14/2/86 "Jerusalem artichokes as a dual purpose crop."

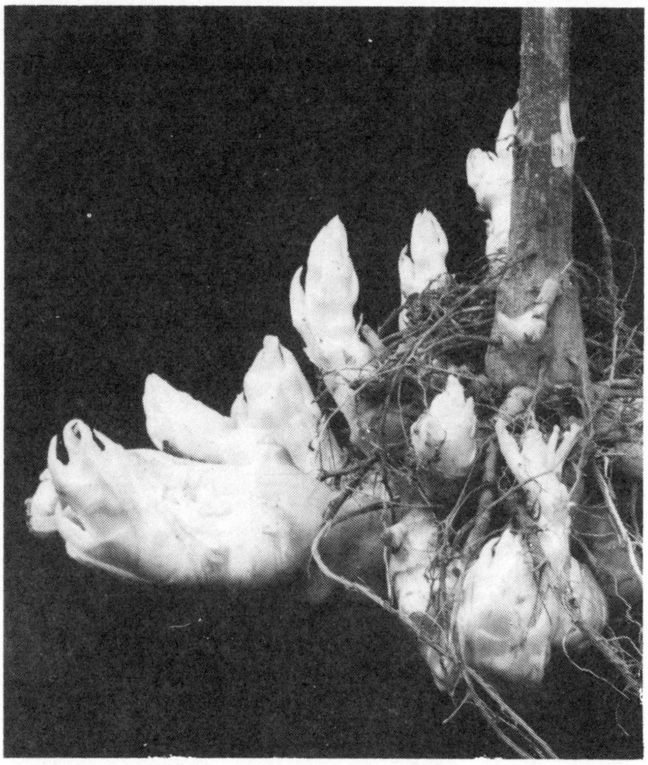

Figure 25: Primitive rhizome development on Jerusalem artichoke seedling.

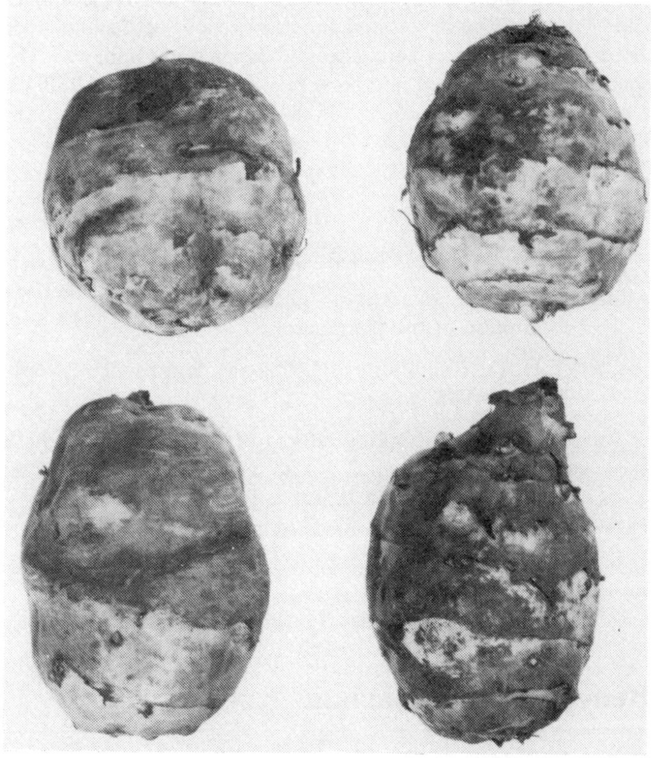

Figure 26: Hybrid of Jerusalem artichoke and the common sunflower (*Helianthus annuus* .)

Figure 27: Kenaf.

Hibiscus cannabinus L.

Malvaceae

Common
Kenaf, Deccan or ambari hemp, Bimlipatan jute.

Synonyms
Hibiscus asper Hook.

Notes
A commercial textile fibre and jute substitute. Widely grown in India. Trials were conducted by MAF in South Auckland in the early 1970's. Plants are annual and reached 2 metres in height at Otara in 1973.

References
Wilson, F.D. and Menzel, M.Y. (1964). Kenaf (*Hibiscus cannabinus*), and roselle (*H. sabdariffa*). Economic Botany 18, 80–91.

Sedgwick, K.N. (1987). Kenaf: can it complement cane. Queensland Agricultural Journal 113(4), 224.

White, G.A. et al. (1970). Cultural and harvesting methods for kenaf. USDA Production Research Report No. 113, Washington. 38 p.

Hibiscus sabdariffa L.

Malvaceae

Common
Roselle, Rosella, Red sorrel, Jamaican sorrel, Hibiscus (N.Z. herbalists), Senegal hemp.

Notes
There are two distinct varieties of this species:
1) var. *sabdariffa* is a bushy annual plant up to 2 m high with fleshy calyces, red coloured in most cultivars, which are used in herbal teas and preserves. There was a small industry making roselle jam at the turn of the century in Queensland and dried calyces are increasingly imported into New Zealand as a source of 'hibiscus tea.'
2) var. *altissima* Wester. Plants are taller, frequently over 4 m in height and are an alternative source of fibre to *H. cannabinus*.

References
Berg, J. (1977). *Hibiscus sabdariffa* (Jamaica sorrel). Alafua 2(2), 5.

Sharaf, A. (1962). The pharmacological characteristics of *Hibiscus sabdariffa* L. Planta Medica 10(1), 48–52.

Crane, J.C. (1949). Roselle, a potentially important plant fibre. Economic Botany 3, 89–103.

Hibiscus tiliaceus L.

Malvaceae

Common
Beach hibiscus, Corkwood, Fau (Samoa and Tonga), Vau (Fiji), 'Au (Rarotonga), Purau (Tahiti), Hau (Hawaii), Burao (New Caledonia).

Synonyms
Pariti tiliaceum (L.) Britton, *Partum tiliaceus* St Hil.

Notes
A small coastal tree (to 15 m) common throughout the tropical South Pacific. The bark fibre is widely used in Polynesia for cordage, kava strainers and the traditional grass skirts. The stems are used for canoe outriggers.

References
Whistler, W.A. (1980). Coastal flowers of the tropical Pacific. Oriental Publishing Co., Honolulu. p 24.

Petard, P. (1986). Quelques plantes utiles de Polynesie Francaise. Editions Haere Po No, Tahiti. pp 214–217.

Figure 28: Roselle.

Degener, O. (1945). Plants of Hawaii National Park illustrative of plants and customs of the South Seas. Edwards Brothers, Ann Arbor. pp 213 – 219.

Hitchenia caulina (Grah.) Baker

Zingiberaceae

Common
Chavar, Arrowroot lily.

Synonyms
Curcuma caulina Grah.

Notes
Native to Western India around Bombay. Cultivated as a source of arrowroot starch, but unimportant (vide *Maranta arundinacea.*) Possible use for erosion control in tropics.

References
Khairnar, M.S. (1945). *Hitchenia caulina* (chavar) as a source of arrowroot. Indian Forester 71(4), 126–127.

Hovenia dulcis Thunb.

Rhamnaceae

Common
Japanese raisin tree, Coral tree, Kempo–nashi (Japan), Zhi ju zi (Pinyin), Sicka (Hindi).

Notes
A tall deciduous tree (to 30 m), with ornamental foliage from eastern Asia. The fleshy peduncles are edible and often sold as dried fruit in the Far East. Trees are rare in northern New Zealand.

References
The Wealth of India, Raw Materials. Vol 5. CSIR, New Delhi 1959. p 133.

Pollock, R.P. and Halliwell, G. (1978). A guide to some lesser known tree and other crops mainly for the warmer parts of New Zealand. MAF, Gisborne. p 14.

Illicium verum Hook.f.

Illiciaceae

Common
Star anise, Chinese anise, Badian.

Synonyms
Badenifera officinarum Ktze.

Notes
A small slender evergreen tree (to 12 m) native to southern China and North Vietnam. The hard, woody, eight–lobed fruit is the spice of commerce used in some liqueurs (e.g. Anisette and Pernod), processed foods and forms 2 to 6% of Japanese curry powders.
N.B. A closely related species with a more northerly range extending to Japan, the shikimi, *I. anisatum* L. (syn. *I. religiosum* Sieb. et Zucc.), has almost identical fruit. These are only readily distinguishable by taste – the latter is not sweet. The principal use of *I. anisatum* is for incense in Buddhist ceremonies.

Figure 29: Star anise capsules.

References

Anon (1888). Star anise. Kew Bulletin No. 19: 173–175.

Rosengarten, F. (1969). The book of spices. Livingston Publishing Co., Wynnewood, Penn. pp 103–105.

Indigofera arrecta Hochst. ex A. Rich.

Leguminosae

Common
Indigo.

Synonyms
Indigofera tinctoria Chev. *var. arrecta* Behr.

Notes
A stout herb, woody towards the base. Usually cultivated as an annual and harvested at the early flowering stage to provide the dye of commerce. The plant is native to tropical Africa but was largely cultivated in the Far East before the introduction of synthetic indigo in 1890.
Two other species were also once of commercial importance; *I. spicata* Forsk., creeping or trailing indigo and *I. tinctoria* L., Indian indigo.
Other species of Indigofera sometimes found in New Zealand gardens include *I. australis* Willd., Australian indigo, *I. gerardiana* from the Himalayas and *I. incarnata* (syn. *I. decora*) a herbaceous species from China.

References

Ghosh, A.K. (1944). Rise and decay of the indigo industry. Science and Culture 9(11), 487–493 and 9(12), 537–542.

Martin–Leake, H. (1975). An historical memoir of the indigo industry of Bihar. Economic Botany 29(4), 361–371.

Inga edulis Martius

Leguminosae

Common
Icecream bean, Pakai (Tuamotu Archipelago), Pati (Austral Is.).

Synonyms
Inga ynga (Vell.) Moore.

Notes
Inga is a large genus of trees from tropical America, many species of which have edible pods. *I. edulis* has been introduced to Hawaii, Queensland, Tahiti and Rarotonga. This may be the species grown in north Auckland (vide Dick Endt, Lansendt Nurseries). The thick four–angled flat or twisted pods contain an edible sweet pulp around the seeds.

References

Anon (1980). *Inga spp.* Annotated bibliography (1931–1980). Queensland Department of Primary Industries, Bibliography Ser. 80–41.

Chapman, K.R. (1984). Leguminosae – *Inga spp.* In P.E. Page (ed). Tropical tree fruits for Australia. Queensland Department of Primary Industries, Brisbane. pp 81–82.

Inocarpus fagifer (Parkinson) Fosberg

Leguminosae

Common
Tahitian chestnut, Polynesian chestnut, Otaheite chestnut, Marau or mape (Tahiti), Ifi (Polynesia), Ivi (Fiji), If or Aila (New Guinea).

Synonyms
Inocarpus edulis Forst., *Boca edulis* (Forst. f.) Baill.

Notes
Tall evergreen trees with fluted trunks and butressed roots in coastal forests of Pacific Islands westwards to Papua New Guinea. Introduced into Taiwan. The pods contain a large single seed which is very edible but under exploited. At one time a staple food for several months of the year in Samoa.

References
Uhe, G. (1974). Wayside plants of the South Pacific. Stocton House, Auckland. pp 125–126.

Henty, E.E. (1982). Some nut bearing plants in Papua New Guinea. Proceedings of the Second Papua New Guinea Food Crop Conference, Port Moresby. pp 78–85.

Duke, J.A. (1989). Handbook of nuts. CRC Press, Boca Raton. pp 175–176.

Petard, P. (1986). Plantes utiles de Polynesie. Editions Haere Po No Tahiti, Papeete. pp 177–178.

Ipomoea aquatica Forsk.

Convolvulaceae

Common
Water convolvulus, Yosai (Japan), Kankong (Malay), Ota kerisi (Fiji).

Synonyms
I. reptans Poir.

Notes
An aquatic herb widely used as a vegetable in Southeast Asia. Less common in the South Pacific.

References
Edie, H.H. and Ho, B.W.C. (1969). *Ipomoea aquatica* as a vegetable crop in Hong Kong. Economic Botany 23, 32–36.

Allen, C.M. (1983). Water spinach – the exotic aquatic. Organic Gardening 30(12), 65–67.

Barrau, J. (1955). *Ipomoea aquatica* , easily grown semi–aquatic vegetable. SPC Quarterly Bulletin 5(3), 19–20.

Jasminum officinale L.

Oleaceae

Common
Poet's jasmine, White jasmine, Italian, Spanish or Catalonian jasmine, Jathimali (Hindi), Su xim hua (Pinyin).

Notes
A climbing shrub native to hill areas from Iran to the Himalayas. Introduced to China at an early date. In New Zealand plants have escaped from cultivation in Marlborough and Canterbury. The larger flowered *Jasminum grandiflorum* L. syn. *J. officinale var. grandiflorum* Bailey, cultivated as a garden plant in northern New Zealand and common in most Pacific Islands is the principal species cultivated in Southern Europe and India for the essential oil which is the 'jasmine' perfume of commerce. The Indian product is a yellowish brown semi–solid substance known as 'jasmine concrete'. Some jasmine oil is also used by the food processing industry as a flavouring, e.g. 'Maraschino cherries'. N.B. *Jasminum sambac* (L.) Ait, Arabian jasmine, is the principal species used in China for 'jasmine tea'. *J. odoratissimum,* a native of Madeira is grown in Taiwan for perfuming tea – 'shuei flowers'.

References
Muthuswamy, S. and Shanmugaelu, K.G. (1982). Jasmine cultivation in southern India. In Atal and Kapur (eds). Cultivation and utilisation of aromatic plants. CSIR, Jamu–Tawi. pp 502–524.

Poucher, W.A. (1979). Perfumes, cosmetics and soaps. 8th edn. Chapman and Hall, London. Vol. 2. pp 111–120.

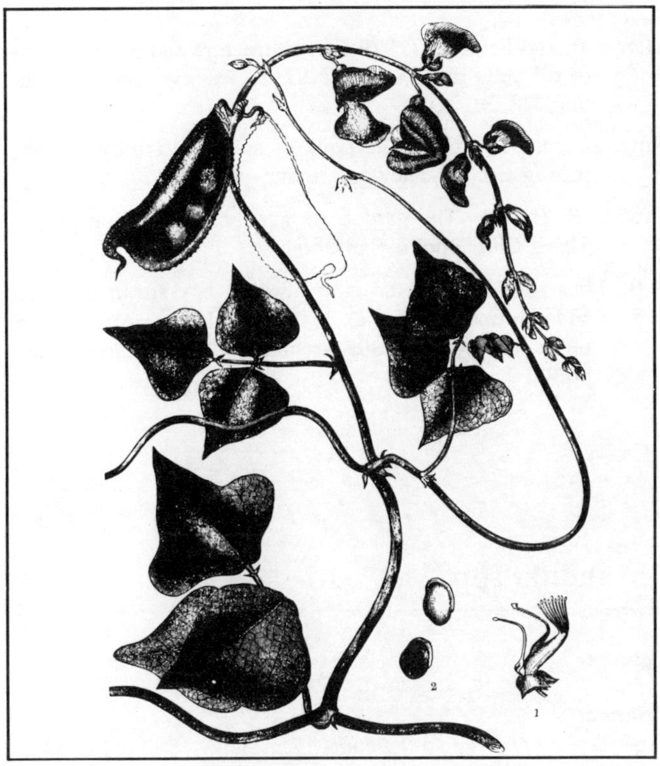

Figure 30: Lablab bean.

Lablab purpureus (L.) Sweet

Leguminosae

Common
Bonavist, Egyptian bean, Hyacinth bean, Lablab bean, Batau (Philippines), Toba (Fiji), Papapa (Hawaii).

Synonyms
Dolichos lablab L., *Lablab niger* Medik.

Notes
A vining bean from tropical Asia, widely cultivated in warm semi–arid areas. Used as a forage crop in Queensland (sacrilege!).

References
National Academy of Sciences (1979). Tropical legumes : resources for the future. NAS, Washington. pp 59–67.

Purseglove, J.W. (1968). Tropical Crops. Dicotyledons. Longman, London. Vol. 1. pp 273–276.

Lagenaria siceraria (Molina) Standley

Cucurbitaceae

Common
Bottle gourd*, Hue (Maori), Vago (Fiji), Ue (Cook Is), Fagufagu (Samoa).The 'penis gourd' of New Guinea is this species (Heiser, 1979).

Synonyms
Lagenaria leucantha (Duch.) Roxb., *Lagenaria vulgaris* Ser.

Notes
A species of unknown origin spread throughout the warm temperate and subtropical regions in prehistoric times. Introduced to New Zealand by the Maori but the strain may have been modified by later introductions (vide Palmer 1985). The immature gourds when eaten in the manner of courgettes are superior in texture.

References
Heiser, C.B. (1979). The gourd book. University of Oklahoma Press, Norman. p 248.

Martin, F.W. (1979). Vegetables for the hot humid tropics. Pt 4, Sponge and bottle gourds. USDA, New Orleans. 19 pp.

See also: Palmer, J. (1985). The gourd *Lagenaria siceraria* (Molina) Standl., as a vegetable. Hue Newsletter No. 3. pp 3,4 & 6. Anthropology Dept. Auckland University.

Wheeler, Jenny (1984). Dotty over bottle gourds. Auckland Star 29/12/84.

*N.B. The calabash gourd, occasionally confused with the bottle gourd comes from a tropical American tree, *Crescentia cujete* L.

Laurus nobilis L.

Lauraceae

Common
True laurel, Sweet bay, Bay tree.

Notes
A small evergreen tree native to the Mediterranean region. Widely cultivated in temperate areas for its spicy leaves which are dried and used in food flavouring. The leaves are sometimes used in veterinary medicine. The closely related *L. canariensis* Webb and Benth., from the Canary Isles and Madeira is used in a similiar manner.

References

Maron, R. and Fahn, A. (1979). Ultrastructure and development of oil cells in *Laurus nobilis* L. leaves. Bot. J. Linn. Soc. 78, 31–40.

Rosengarten, F. (1969). The book of spices. Livingston Publishing Co., Wynnewood, Penn. pp 118–127.

Creasy, R. (1986). The gardener's handbook of edible plants. Sierra Club Books, San Francisco. pp 276–278.

N.B. Most references to the cultivation of bay are in Russian. Vide Simon, J.E. et al. (1984). Herbs. An indexed bibliography, 1971–80. Archon Books, Hamden, Conn. pp 9–10.

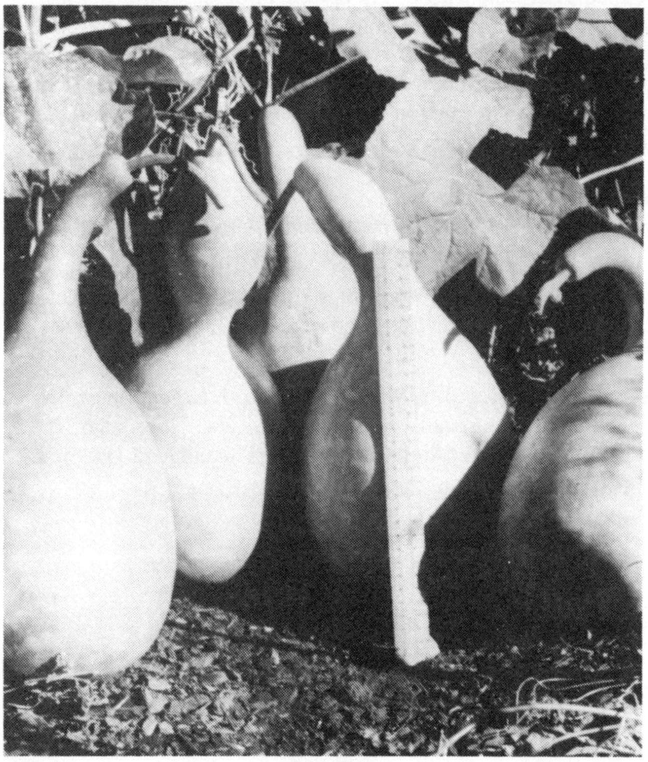

Figure 31: 'Maori' gourds in a Christchurch garden, March 1983.

Lavandula spp.

Labiatae

Common

True lavender, English lavender, Lavandin, Spike lavender, Spanish or topped lavender.

Notes

Major species cultivated in New Zealand include:
Lavandula angustifolia Mill. (syn *L. officinalis* Chaich., *L. spica* Lois. and *L. vera* DC.) True lavender, English lavender.
L. x intermedia Emeric. non Lois., Lavandin.
L. latifolia Medicus (syn *L. spica* DC.) Spike lavender.
L. stoechas L. French or topped lavender.
L. dentata Lois., toothed lavender.
There are many cultivars of *L. angustifolia* and *L. x intermedia* in N.Z. In *L. angustifolia,* Tucker(1985) includes: Alba, Nana alba, Rosea, Dwarf blue, Hidcote, Lodden blue, Middachten, Nana, Atropurpurea, Mitcham grey and Munstead.
In *L. x intermedia* Alba, Dutch, Grosso, Hidcote giant, Old English, Provence, Seal, and Silver gray.
L. stoechas has been a weed of pasture land in Victoria since 1920.

References

Tucker, A.O. (1985). The cultivars of lavender and lavandin. Baileya 22(4), 168–177.

Lammerink, J. (1988). Grosso lavandin for oil. CRD Agricultural Bulletin No 11. DSIR, Christchurch. 16 pp.

See also: Guenther, E. (1954). The French lavender and lavandin industry. Economic Botany 8, 166–173.

Lawsonia inermis L.

Lythraceae

Common
Henna, Mignonette bush, Mehndi or Mendhi (Hindi), Kiari (Fiji).

Synonyms
Lawsonia alba Lam.

Notes
A deciduous shrub with sweet smelling flowers, indigeous to a large area from North Africa to Western India. Now introduced throughout the tropics, including many Pacific Islands, as an ornamental or hedge plant. The use of dried henna leaves as a cosmetic orange dye is traditional in India and Egypt. Plant parts are also employed in native medicine and turn up in 'health food' shops in New Zealand.

References
The Wealth of India, Raw Materials. Vol 6. CSIR, New Delhi 1962. pp 47–50.

Burkill, I.H. (1935). Economic products of the Malay Peninsula. Crown Agents, London. pp 1323–1325.

Lepidium meyenii Walp.

Cruciferae

Common
Maca.

Notes
An Andean root crop. Not known in N.Z. but would probably grow here.

References
Leon, J. (1964). The maca (*Lepidium meyenii,*) a little known food plant of Peru. Economic Botany 18(2), 122–127.

Kay, D.E. (1973). TPI Crop and Product Digest No. 2. Root crops. Tropical Products Institute, London. pp 93–95.

Figure 32: Curled cress plants.

Lepidium sativum L.

Cruciferae

Common
Garden cress, Peppergrass.

Synonyms
Nasturtium crispum Bauh.

Notes
An annual herb, cultivated in Europe for centuries as a salad crop. Often used at the seedling stage (e.g. mustard and cress) but if allowed to grow on provides useful salading from autumn to early spring when it normally flowers. There are two basic types of cultivar; those with curled leaves (cf. parsley) and 'broad leafed' cress. The former is less hardy and more easily damaged by inclement weather but could be commercially exploited as a garnish.

References
Kennelly, A.G. (1952). The home vegetable garden. N.Z. Department of Agriculture Bulletin 243, Wellington. p 102.

Limnanthes alba Benth.

Limnanthaceae

Common
Meadowfoam.

Notes
Limnanthes spp. are native to damp meadow areas of Pacific North America. Several species have seed oils with a high percentage of acids with more than 18 carbon atoms. Most agronomic work on *L. alba* has been carried out at Oregon State University, Corvallis. Seed was introduced to New Zealand in 1977 but initial test plots at Lincoln were unharvestable due to the low stature of the plant and weed infestation. Later trials have been undertaken at Massey University (Dr Ian Gordon) specifically with the cultivar 'Foamore' (vide Commercial Horticulture, June 1983 p 12).
N.B. *Limnanthes douglasii* R. Br. is often grown as an ornamental annual plant under the common name of 'poached egg flower'.

References
Jolliff, G.D. (1981). Development and production of meadowfoam. In: E.H. Pryde et al. (eds) 'New sources of fats and oils', AOCS Monograph No. 9, pp 269–285. Reissued as Oregon State Agricultural Experiment Station Technical paper No. 5622.

Calhoun, W. and Crane, J.M. (1978). Seed yields of meadowfoam as influenced by N, seeding rates, and soil–water table levels. Agronomy Journal 70, 924–926.

Lippia citriodora HBK.

Verbenaceae

Common
Lemon verbena.

Synonyms
Aloysia citriodora Ort., *Aloysia triphylla* Britton, *Verbena triphylla* L'Her.

Notes
A shrubby plant with lemon scented leaves, native to Argentina and Chile, but widely cultivated in warm temperate and sub–tropical gardens. It is marginally hardy in Christchurch. Verbena oil distilled from the leaves has been largely replaced in perfumery by lemon grass and other cheaper sources.

References
Morton, Julia (1976). Herbs and spices. Golden Press, New York. p 42.

El–Hamidi, A. et al. (1983). *Lippia citriodora* grown in Egypt. A new crop under development. Acta Horticulturae 132, 31–33.

Litchi chinensis Sonn.

Sapindaceae

Common
Lychee, Litchi.

Synonyms
Euphoria litchi Desf., *E. sinensis* Gmel., *Litchi litchi* Briton, *Nephelium chinense* Druce., *N. litchi* Camb.

Notes
A fruit tree from southern China widely introduced and cultivated in sub–tropical areas elsewhere in the world including India, Taiwan, South Africa and Queensland, but not in the tropics e.g. Malaysia. Litchi have been introduced to many Pacific Islands and have prospered in New Caledonia but remained neglected elsewhere until recently when commercial plantings began in the Cook Islands. There are also small scale plantings in northern New Zealand.

References
Cull, B.W. and Paxton, B.F. (1982). The litchi. Australian Horticulture 80(12), 87–99.

Menzel, C.M. et al. (1986). The lychee in Australia. Queensland Agricultural Journal 114(1), 19–28.

Luffa acutangula (L.) Roxb.

Cucurbitaceae

Common
Angled loofah, Ridged gourd, Silk gourd, Szekwa (Chinese), Kali tori (Hindi), Seequa (Hawaii).

Synonyms
Cucumis acutangula L. *Trichosanthes amara,* Blanco.

Notes
An edible gourd probably originating in India but widely cultivated in southern China and many Pacific Islands.

References

Porterfield, W.M. (1943). Luffas as they are used by the Chinese. Journal of New York Botanic Garden 44, 134–138.

Martin, F.W. (1979). Vegetables for the hot humid tropics, Pt 4. Sponge and bottle gourds, *Luffa* and *Lagenaria* . USDA, New Orleans. 19 pp.

Luffa cylindrica (L.) M. Roemer

Cucurbitaceae

Common
Loofah, Luffa, Dishcloth gourd, Vegetable sponge, Shuikwa (Chinese), Ghiya tori (Hindi), Vusovuso (Fiji), Motini (Cook Is), Aioro (Tahiti).

Synonyms
Luffa aegyptica, Mill., *Momordica cylindrica* L.

Notes
The immature fruit are edible but the species is principally known for the fibrous interior of the ripe gourds. Probably originated in India but widely cultivated in China, Japan and many Pacific Islands.

References

Porterfield, W.M. (1955). Loofah – the sponge gourd. Economic Botany 9, 211–223.

Howes, F.N. (1931). The loofah industry. Kew Bulletin (4), 266–270.

Lupinus albus L.

Leguminosae

Common
White lupin, Term (Egypt).

Synonyms
Lupinus termis Forsk.

Figure 33: White lupin with edible pods.

Notes
An ancient food crop in the Mediterranean area. Still important in parts of the Middle East and less so in Italy for human consumption. Some modern low alkaloid cultivars such as 'Ultra' (Pfluggs, Germany) are particularly useful as the pods may be sliced in a similiar manner to green beans.

References

Gladstones, J.S. (1970) Lupins as crop plants. Field Crop Abstracts 23(2), 123–148

Williams, W. (1984). Lupins in crop production. Outlook on Agriculture 13(2), 69–76.

See also: Palmer, J. (1975). Lupins for South Auckland. New Zealand Farmer, February 27th 1975, p 45.

Lupinus mutabilis Sweet

Leguminosae

Common
Pearl lupins, Tarwi (South America), Altramuz (Spanish).

Synonyms
Lupinus tauris Hook.

Notes
A legume seed crop little known outside the Andean regions until recently. Cultivars are currently undeveloped producing indeterminate plants ripening seed over a prolonged period, and plants usually adapted to a long growing season of at least 5 months. Low alkaloid strains have been reported but are not yet available as cultivars. Some trials of pearl lupins have been conducted at Lincoln College (vide Dr G. Hill).

References
National Academy of Sciences (1979). Tropical legumes resources for the future. NAS, Washington. pp 86–91.

Torres–Tello, F. (1976). *Lupinus mutabilis* Sweet – a potent food source from the Andean region. The American Journal of Clinical Nutrition 29, 933.

Malpighia punicifolia L.

Malpighiaceae

Common
West Indian cherry, Barbados cherry, Native cherry, French cherry, Acerola (Spanish), Malpi (Philippines).

Synonyms
Malpighia glabra Millsp. non L.

Notes
A small evergreen fruit tree (to 6 m), native to the West Indies but planted commercially in Queensland, Florida, Venezuela, India, Philippines and Hawaii. During the 1960's the fruit was widely promoted as a source of Vitamin C but did not compete with the cheaper synthetic product. The species is rare or absent from most islands in the South Pacific but deserves to be better known, if only for its ornamental value.

References
Ostendorf, F.W. (1963). The West Indian cherry. Tropical Abstracts 18(3), 145–150.

Morton, Julia F. (1987). Fruits of warm climates. Creative Resource Systems, Winterville, N.C. pp 204–207.

See also: Anon (1980). *Malpighia spp.* Annotated bibliography (1969–1980). Queensland Department of Primary Industry, Bibliography Series 80–37. 18 pp.

Malva verticillata L.

Malvaceae

Common
Whorled mallow, Cheesewood, Dong kui zi (Pinyin), Fuyn aoi (Japan). The 'mauve frisee' of France is *M. verticillata var crispa* Makino.

Synonyms
Malva crispa L., *M. mohileviensis* Graebn., *M. palmiroalaica* Ilj.

Notes
This species was domesticated in China at an early date, and was an important vegetable of the Tang era (7th to 10th centuries AD) but has since been replaced by Chinese cabbage (Li, 1969). It is still used as a vegetable in the hill areas of Northeastern India and a curled leafed cultivar is used for garnishing in France. The whole plant and also the seeds of this species are included in the Chinese Pharmacopoeia (1977) as a mild cathartic and for nursing mothers.

References
Li Hui–lin (1969). The vegetables of ancient China. Economic Botany 23, 253–260.

Singh, H.B. and Arora, R.K. (1978). Wild edible plants of India. Council of Agricultural Research, New Delhi. p 28.

N.B. *M. rotundifolia* L., dwarf mallow, is an ancient vegetable crop in the eastern Mediterranean. It was used in ancient Greece and Rome and is still cultivated to a limited extent in Egypt.

vide: Duke, J.A. (1985). Handbook of medicinal herbs. CRC Press, Boca Raton. p 291.

Neither of the above species are known in New Zealand. Plants sometimes recorded as *M. rotundifolia* are *M. neglecta* Wallr. also known as dwarf mallow. One specimen of *M. verticillata* was obtained from Christchurch late last century but has not been found since.

Mangifera indica L.

Anacardiaceae

Common
Mango, Manua (Samoa), Vi papa'a (Cook Is.), Tumu vi (Tahiti).

Notes
Large tree from tropical eastern Asia. Needs dry season for flowering and fruit set. Widespread in South Pacific Islands but of little commercial importance except in the Philippines.

References
Iqbal, M. (1982). Review of mango research in Fiji. Agricultural Journal, Fiji 44(1), 21–26.

Whiley, A.W. (1984). Mango. In P.E. Page (ed.), Tropical tree fruits for Australia. Queensland Department of Primary Industries, Brisbane. pp 25–31.

Manihot esculenta Crantz

Euphorbiaceae

Common
Cassava, Manioc, Tapioca.

Synonyms
Manihot utilissima Pohl

Notes
A starchy root crop which keeps well until dug and is propagated by stem cuttings. Of South American origin but now a major tropical root crop worldwide due to the ease of cultivation.

References
Very numerous!

Cock, J.H. (1985). Cassava: new potential for a neglected crop. Westview Press, London. 190 pp.

Kay, D.E. (1973). TPI Crop and Produce Digest 2. Root crops. Tropical Products Institute, London. pp 24–39. (Lists 53 references).

See also: Thaman, R.R. and Thomas, P.M. (1982). The cassava invasion: the cultural, nutritional and ecological impact of cassava on Pacific Island food systems. Proceedings of Second Papua New Guinea Food Crop Conference. Port Moresby. pp 330–349.

Maranta arundinacea L.

Marantaceae

Common
Arrowroot*, West Indian arrowroot, Kuzu–ukon (Japan).

Notes
A herbaceous perennial with long starchy rhizomes. Native to tropical America and cultivated in the West Indies. Widely introduced through the tropical Pacific but not known to be cultivated on any scale.

References
Erdman, M.D. and Erdman, B.A. (1984). Arrowroot, food, feed, fuel and fibre resource. Economic Botany 38(3), 332–341.

Erdman, M.D., Photak, S.C. and Hall, H.S. (1985). Potential for production of arrowroot in the southern United States. J. Amer. Soc. Hort. Sci. 110(3), 403–406.

*N.B. other 'arrowroots': –
Indian arrowroot or Shoti – *Curcuma zedoaria* .
Arrowroot lily or Chavar – *Hitchenia caulina* .
East Indian arrowroot – *Tacca leontopetaloides* .
Queensland arrowroot– *Canna edulis* .

Melaleuca alternifolia (Maiden and Betche) Cheel

Myrtaceae

Myrtaceae

Common
Tea tree (in Australia), Narrow leaved paperbark tea tree.

Synonyms
Melaleuca linariifolia Smith *var. alternifolia* Maiden and Betche.

Notes
A report on the medical and dental uses of 'Ti–trol' (the commercial name for the oil) was published in 1936 by Australian Essential Oil Ltd. Since then the use of this skin penetrating bacteriostat has waxed and waned. According to Lassak and McCarthy (1983) from whom the above information was derived, the eventual decline of tea tree oil was due to unreliable supply, inconsistent quality and lack of promotion. However, another Australian company began to show interest in 1984 and enquiries were made in N.Z.
M. alternifolia has been grown in New Zealand as an ornamental (?), but the swampy ground favoured by this species makes harvesting difficult and the species itself is markedly variable.

References

Lassak, E.V. and McCarthy, T. (1983). Australian medicinal plants. Methuen Australia, North Ryde. pp. 95–97.

Small, B.E.J. (1982). Tea tree oil. Department of Agriculture, NSW, Agfact P6.2.1. 4 pp.

N.B. Another Australian native *M. quinquenervia* (Cav.) St Blake is a source of 'niaouli oil'.

vide: Morrison, F.R. (1958). Essential oil from tea–trees. S.P.C. Quarterly Bulletin 8, 47–49.

Neither of these Australian species are as valuable as 'cajeput' oil from *M. cajuputi*.

Melaleuca cajuputi Powell.

Myrtaceae

Common

Cajeput* (from the Malay; Kayu – a tree, Puteh – white), Paperbark tree.

Synonyms

Melaleuca caja–putti Hort. ex DC., *M. leucadendron* of some authors but not the Australian species. Vide: S.T. Blake(1968). A revision of *Melaleuca leucadendron* and its allies. Contributions from the Queensland Herbarium I, 1-114. The synonyms are too numerous to list here.

Notes

A small evergreen tree with a wide distribution from Northern Australia through Indonesia and Malaysia to Indochina and India. Since it has been cultivated for centuries it is doubtfully native throughout this range. The oil distilled from the young shoots and foliage is a traditional medicine in Southeast Asia and has been used in British medicine since 1809. The 1963 edition of the British Pharmaceutical Codex suggests cajeput oil as a counter irritant in ointments and liniments but in the past its prescription has been far wider; from arthritis to toothache. It is a major constituent of 'Tiger balm'.

References

Burkill, I.H. (1935). Economic products of the Malay Peninsula. Crown Agents, London. Vol. 2. pp 1431–1433.

Quisumbing, E. (1978). Medicinal plants of the Philippines. Katha Publishing, Quezon City. pp 663–664.

*Cajeput in the U.S.A. apparently refers to *Melaleuca quinquenervia* (Cav.) S.T. Blake. Vide: Morton, Julia F. (1966). The cajeput tree – a boon and an affliction. Economic Botany 20, 30–39. The oil from this species is marketed in New Caledonia as 'Niaouli oil'. Vide: Morrison, F.R. (1958). Essential oils from 'tea–trees'. SPC Quarterly Bulletin 8, 47–49.

For the latest review of the nomenclature of Melaleuca spp. Vide: Byrnes, N.B. (1986). A revision of *Melaleuca* L. in northern and eastern Australia. Austrobaileya 2(3), 254–273.

Mentha x piperita L.

Labiatae.

Common

Peppermint, Bergamot mint.

Notes

A variable hybrid plant widespread in damp places on both islands of New Zealand. A cultivar suitable for commercial extraction of peppermint oil – 'Mitcham', was introduced to New Zealand in 1968. Subsequent commercial production was hindered by mint rust (Puccinia menthae) and oil quality.

N.B. Japanese peppermint is *M. canadensis* L. (syn. *M. arvensis f piperascens* Malinv. ex Holmes).

References

Clark, R.J. and Menary, R.C. (1981). Variation in composition of peppermint oil in relation to production areas. Economic Botany 35(1), 59–69.

Lammerink, J. et al. (1977). Peppermint growing in Canterbury. Lincoln College, Canterbury. 38pp.

Metroxylon sagu Rottb.

Palmae

Common
Sago palm, Rembia (Malay), Lumbia (Philippines), Sak–sak (New Guinea), Sogo (Fiji), Bia (Solomon Is), Niu masoa (Samoa),

Synonyms
Metroxylon inerme Mort., *Sagus inermis* Roxb., *Sagus rumphii* Ait.

Notes
A palm with a short, thick, starch–filled trunk, flowering terminally. Sago starch is usually extracted in the immediate pre–flowering stage and is an important food source in coastal swamp areas of New Guinea and Melanesia, but not in Polynesia. There related species, *M. vitiense* (Wendl.) ex Hook.f. in Fiji, *M. salomense* (Warb.) Becc. in Samoa are only used for thatching.
M. amicarum (Wendl.) Becc. is the ivory nut from the Caroline Islands. The solid seeds are used for button manufacture in Hawaii.

References
Barrau, J. (1959). The sago palm and other food plants of marsh dwellers in the South Pacific. Economic Botany 13(2), 151–162.

Stanton, W. and Flach, M. (eds) (1981) Sago – the equatorial swamp as a natural resource. First International Sago Symposium, Malaysia 1979. Martius Nijhoff, Amsterdam. 244 pp.

See also: Ruddle, K. et al (1978). Palm sago: a tropical starch from many lands. Australian National University Press, Canberra. 207 pp.

Momordica charantia L.

Cucurbitaceae

Common
Bitter gourd, Balsam pear, Peria (Malay), Kiari (Japan), Fu kwa (Chinese), Vaine initia (Tonga), Kerala (Fiji Indian), Pupuruvi (Tahiti).

Synonyms
Momordica indica Stickm., *M. elegans* Salisb.

Notes
A scandent herb with small, distinctively knobbly fruit, bitter tasting but appreciated in Southeast Asia and by Indian migrants to the Pacific. A widespread adventive and possibly native to coastal sites in many Pacific Islands. It was found in Tahiti on Cook's first visit. Young fruits and shoots are used as a vegetable. The fruits are also medicinal.
Related species: *M. dioica* L. is cultivated in Southern India and Sri Lanka (Tampai in Tamil) and *M. cochinchinensis* (Lour.) Spreng., the spiny bitter cucumber, in India, Japan and New Guinea.

References
Walters, T.W. and Decker–Walters, Deena, S. (1988). Balsam pear (*Momordica charantia* , Cucurbitaceae). Economic Botany 42(2), 286–288.

Herklots, G.A.C. (1972). Vegetables in S.E. Asia. Allen & Unwin, London. pp 333–338.

Anon (1962). Wealth of India – Raw materials Vol. 6. CSIR, New Dehli. pp 408–411.

Morinda citrifolia L.

Rubiaceae

Common
Awl tree, Indian mulberry, Kura (Fiji), Nono or noni (Polynesia).

Notes
A widespread tree occuring naturally from the islands of the South Pacific to Southeast Asia. The fruit was at one time of some importance in the diet of Pacific Islanders. But the taste is insipid and little use of this plant is made today except as a source of red dye from the bark or yellow from the roots. This latter dye called 'al' in India is widely used in that region.

References
The wealth of India, Raw Materials Vol 6. CSIR, New Delhi, 1962. pp 423–424.

Uhe, G. (1974). Wayside plants of the South Pacific. Stocton House, Auckland. pp 119–120.

Petard, P. (1986). Plantes utiles de Polynesie. Editions Haere Po No Tahiti. pp 280–286.

Degener, O. (1945). Plants of Hawaii National Park illustrative of plants and customs of the South Seas. Edwards Brothers, Ann Arbor. pp 282 – 286.

Moringa oleifera Lam.

Moringaceae

Common
Drumstick tree, Horseradish tree, Ben nut*, Mungna (Hindi), Malungai (Philippines), Sajinda (Sanskrit but used in Fiji).

Synonyms
Moringa pterygosperma Gaertn.

Notes
A small tree (to 10 m) native to northwestern India but cultivated throughout the sub–continent for its long cylindrical seed pods which are a favoured vegetable. The 'drumsticks' are grown by the Indian population of Fiji and are occasionally imported into N.Z. vegetable markets. They are also produced in the Philippines. The seed oil (ben oil) is used in medicine and to a limited extent for edible purposes and scent. The roots are the portion said to have the flavour of horseradish.
*The 'ban' or 'ben' of Egypt is *Moringa aptera* Gaertn. Both species produce 'ben oil'.

References

The wealth of India, Raw Materials, Vol. 6. CSIR, New Delhi, 1962. pp 426–429.

Peter, K.V. (1979). Drumstick, a multipurpose vegetable. Indian Horticulture 23(4), 17–18.

Ramachandran C. et al.(1980). Drumstick (Moringa oleifera): A multipurpose Indian vegetable. Economic Botany 34(3), 276-283.

Duke, J.A. (1989). Handbook of nuts. CRC Press, Boca Raton. pp 214–217.

Anon, (1904). The nature and commercial uses of ben oil. Bulletin of the Imperial Institute 2, 117–120. (Unsighted).

N.B. For medicinal uses see:
Quisumbing, E. (1978). Medicinal plants of the Philippines. Katha Publishing Co., Quezon City. pp 346–349.

Muntingia calabura L.

Eleocarpaceae

Common
Jamaica cherry, Panama berry, Capulin*.

Notes
A small tree (to 12 m) with spreading branches. Native to tropical America. Often planted in S.E. Asian gardens as a shade tree. Introduced to Tahiti and the Tuamotu Islands where it has proved adaptable to the calcareous sands of coral islands. The red or yellow cherry like fruit containing many small seeds are very sweet.
*N.B. Capulin cherry is strictly speaking *Prunus salicifolia.*

References

Morton, Julia F. (1987) Fruits of warm climates. Creative Resource Systems, Winterville. pp 278 – 280.

Neal, Marie C. (1965). In the gardens of Hawaii. Bernice P. Bishop Museum, Special Publication 50, Honolulu. pp 543 – 544.

Murraya koenigii (L.) Spreng.

Rutaceae

Common
Curry leaf, Kathnim or Kath (Hindi), Oba gekkitsu (Japan), Tej patti (Fiji).

Notes
A small tree (to 5m) native to India with aromatic leaves used as flavouring. Introduced by Indian migrants to the Pacific. The related *M. paniculata* (L.) Jack has an extended natural range through New Guinea and northeastern Australia to New Caledonia. This species the, 'orange jessamine' or 'Chinese myrtle' is often planted in Pacific Island gardens as an ornamental or hedge plant and produces a valuable timber.

References

Philp, J. (1981). Curry leaf and its uses. World Crops 33(6), 125–126.

Peter, K.V. (1978). Curry leaf – a good ingredient for vegetable preparations. India Farmer's Digest 10(12), 13–14.

Musa textilis Nee

Musaceae

Common
Abaca, Manilla hemp.

Notes
A tall slender diploid banana species, domesticated in and possibly native to the Philippines. The outer sheathing leaves of the pseudostems contain strong fibres which produce the Manilla hemp of commerce. This is used for marine ropes and in the fishing industry. Low grade fibres are used for such modern junk food appendages as tea bags. Major production areas are still centred in the Philippines but some abaca has been grown in Malaysia, Sumatra and New Guinea. Production commenced in central America during World War 2 but this was discontinued in 1960. The fruit is fertile and inedible.
A hybrid between *M. textilis* and *M. balbisiana,* an edible diploid, is the source of 'Canton fibre'.

References
Spencer, J.E. (1953). The abaca plant and its fibre. Economic Botany 7, 195–213.

Purseglove, J.W. (1972). Tropical crops – Monocotyledons. Longman, London. Vol 2, pp 377–384.

N.B. For basic peasant agronomy see:
Copeland, E.B. (1911). Abaca. Philippines Agriculturist 1(4), 64–73.

Musa troglodytarum L.

Musaceae

Common
Fehi (Hawaii), Fe'i (Tahiti), Soaqa (pronounced soanga – Fiji), Soa'a (Samoa), Uatu (Rarotonga), N'dare (New Caledonia).

Synonyms
Musa fehi Bertero, *M. uranoscopus* Rumph.

Notes
Possibly native to New Guinea and spread along migration routes into the Pacific as far as Hawaii. At one time an important food source on Tahiti and Fiji. Now naturalised in hill forest areas of many Pacific Islands. The species is a fertile diploid but the short thick orange skinned fruit are frequently seedless, especially when cultivated at low altitudes. The fruits contain a deep yellow juice which discolours the urine of its consumers. The blackish pseudostems have a purple coloured sap. Tradition has it that the first bibles produced in Tahiti were printed with ink prepared from this sap.

References
MacDaniels, L.H. (1947). A study of fe'i banana and its distribution with reference to Polynesian migration. P. Bernice Bishop Museum Bulletin 190, Honolulu. 66 pp.

Cheesman, E.E. (1949). Classification of bananas, 3 – Critical notes on Musa fehi. Kew Bulletin 1949, Pt 4, 445–449.

N.B. Major commercial cultivars of banana are derived from polyploid hybrids of *Musa acuminata x balbisiana.* For a well illustrated review of both diploid and polyploid types see:

Bourke, R.M. (1976). Notes on types of bananas. Proceedings of the 1975 Papua New Guinea Food Crops Conference. Department of Primary Industry, Port Moresby. pp 371–377.

Myristica fragrans Houtt.

Myristicaceae

Common
Nutmeg, Atoge papalagi (Samoa).

Notes
A tree with spreading branches from the Moluccas in Indonesia. Has been planted in the Pacific, particularly in Fiji, but not on a commercial basis. The seed (nutmeg) is surrounded by a red fleshy aril (the mace).
Several other species of Myristica are native to the South Pacific area but are of no commercial importance. Vide: A C Smith, Flora Vitiensis Nova Vol. 2. Lawai, Hawaii 1982. pp391–397.

References
Purseglove, J.W. et al. (1981). Spices. Longman, London. pp 174–228.

Weil, A.T. (1965). Nutmeg as a narcotic. Economic Botany 19, 194–217.

Nelumbo nucifera Gaertn.

Nymphaeaceae

Common
Lotus, Sacred lotus, Hasu (Japan).

Synonyms
Nelumbium speciosum Willd., *Nelumbium nelumbo* Druce.

Notes
An oriental water lily widely grown as an ornamental pond plant in warm temperate and sub-tropical regions. The rhizomes, seeds and petioles are utilised as a vegetable in the Far East and to a limited extent in Tahiti and Hawaii. Grows well as an ornamental water plant in northern N.Z. but rhizome and seed production are poor.

References
Irvine, F.R. and Trickett, R.S. (1953). Water lilies as food. Kew Bulletin (3), 363–370.

Kay, D.E. (1973) TPI Crop and Product Digest No. 2 Root crops. Tropical Products Institute, London. pp 89–92.

Creasy, R. (1986). The gardener's handbook of edible plants. Sierra Club Books, San Francisco. pp 156–159.

N.B. For an historical treatise see also:– Swarap, V. (1983) Lotus. Indian Horticulture 27(4), 2–3

Nigella sativa L.

Ranunculaceae

Common
Black cumin, Small fennel, Nutmeg flower, Roman coriander, Kalonji (Hindi), Jintan hitam (Malay).

Notes
Small annual plants cultivated as a spice seed in southern Europe, the Middle East, USSR, northeastern Africa and India. Black cumin seed is a common addition to bread in these areas and to curry powders in India. Though adapted to warm, arid climates, black cumin grows quite well in the Auckland area. There are potential markets in the Middle East.

References
Jansen, P.C.M. (1981). Spices, condiments and medicinal plants of Ethiopia. Centre for Agricultural Publication and Documentation, Wageningen. pp 75–85.

The Wealth of India – Raw Materials. Vol. 7. CSIR, New Delhi 1966. pp 63–65.

Oenanthe javanica DC.

Umbelliferae

Common
Water dropwort, Seri (Japan), Shelum or selom (Malay), Tespong (Indonesia), Pak an (Thai).

Synonyms
Dasiloma japonicum Miq., *Oenanthe linearis* DC., *O. stolonifera* Wall.

Notes
A perennial herb spreading by stolons, usually growing on a pond or stream margin. *Oenanthe javanica* is found in and is possibly native to a very wide area from northern Japan through China and Southeast Asia to Queensland. In Indonesia and Japan it is a widely used culinary herb and cultivated for this purpose. Elsewhere in Southeast Asia its use is more occasional.

References
Ochse, J.J. (1977). Vegetables of the Dutch East Indies. Australian National University Press, Canberra. pp 717–719.

Yashiroda, K. (1968). A dictionary of popular Japanese herbs and their uses today. Plants and Gardens 24(2), 44–58.

Oenothera biennis L.

Onagraceae

Common
Evening primrose, German rampion.

Notes
A biennial herb usually grown as an annual. Naturalised in New Zealand in the 19th century and cultivated as an oilseed particularly in Canterbury since 1980. The seed contains an essential fatty acid, gamma linolenic (GLA) which is a reported prostaglandin precursor. The trade name of 'Efamol' has been registered with the British Medical Association. In the 19th century the roots of this species were occasionally eaten as a vegetable in Europe. The flavour is bland and the small amount of oil in the roots contains no GLA.

References
Palmer, J. (1985). The evening primrose – a source of essential fatty acid. New Zealand Agricultural Science 19(3), 119–121.

Cutting, O. (1984). Evening primrose – a crop with a future? Arable Farming 11(3), 34 and 40.

Figure 34: Evening primrose.

For medical aspects see:

Horrobin, D.F. (ed.). Clinical uses of essential fatty acids. Eden Press, Montreal and London, 1982. 237 pp.

Notes

A small evergreen tree from the eastern Mediterranean and cultivated there for several millennia. Now also grown in California, southeastern Australia and South Africa. Most trees in New Zealand are seedlings of no commercial value but some named cultivars have been around for many years and a small commercial planting has commenced in Marlborough (Clark, 1988). Unfortunately the small collection at the DSIR Te Puke orchard was pulled out in 1982. The major use of olives is as a source of edible oil extracted from the pericarp of the fruit. To obtain a high oil content in the pericarp, fruit require periods of 6 months or more growth before they are harvested. Some winter chilling is necessary for flowering and cross pollination is essential. Otherwise olive trees are undemanding. Any New Zealand production of olives will probably be used for pickled fruit only.

References

Clark, R. (1988). Growing the venerable olive. N.Z. Gardener 44(12), 57–59.

Hartmann, H.T. (1962). Olive growing in Australia. Economic Botany 16(1), 31–44.

Garoyan, L. and Horel, L. (1980). California olives: situation and outlook. California Agriculture 34(2), 10–12.

See also: Anon (1975). Home pickling of olives. Division of Agricultural Sciences, University of California. Leaflet 2758. 10 pp.

Anon, (1944). Tropical and sub–tropical fruits in the Gisborne district. N.Z. Journal of Agriculture 68, 277–278. (2 cultivars – Sevillano and Mission.)

Morgan, W. (1891). Olive culture. The N.Z. Country Journal, 15(4), 328–331.

Olea europaea L.

Oleaceae

Common
Olive.

Synonyms
Olea sativa Hoffsmeg. et Link.

Origanum dictamnus L.

Labiatae

Common
Dittany of Crete, Hop marjoram.

Synonyms
Amaracus dictamnus Benth.

Notes

A low growing tender perennial plant which is restricted in New Zealand by frosts in the south and wet winters in the north. It is used in Greek herbal teas with names like 'dictamos' and 'erontas' and has some medicinal value.

Other Origanum spp. grown in New Zealand as culinary herbs include: *Origanum majorana* L. Sweet marjoram and *O. vulgare* L. Wild marjoram. A cultivar of this last species – cv. Viride, is the oregano of Italy. Vide: Painter, G. (1985). Herbs for every purpose. Copy–type Office Services, Glen Eden. 21 p.

Greek oregano, usually assumed to be the most desirable species is *O. vulgare ssp hirtum* (Link.) Ietswaart. Vide: Fleisher(1988), below.

References

Skrubis, B. (1979). *Origanum dictamnus,* a Greek native plant. Journal of Ethnopharmacology 1, 411–415.

Harvala, C. et al (1987). Essential oil from *Origanum dictamnus* L. Planta Medica 53(1), 107–109.

See also: Fleisher, A. and Fleisher, Z. (1988). Identification of Biblical hyssop and origin of the traditional use of oregano–group herbs in the Mediterranean region. Economic Botany 42(2), 232–241.

Ietswaart, J.H. (1980). A taxonomic revision of the genus *Origanum* (Labiatae). Lieden University Press, The Hague. (unsighted).

Figure 35: Dittany of Crete.

Oxalis tuberosa Molina

Oxalidaceae

Common

Oca. Sometimes mistakenly called 'yam' in New Zealand.

Notes

A small Andean tuber with a range of skin colours. Only two pink cultivars are common in N.Z. Many native Andean cultivars contain appreciable quantities of oxalic acid. Oca are short–day plants and tubers are not produced on the ends of axillary branches until late autumn. The most recent accessions were imported from Universidad National de Loja, Ecuador (Francisco Castro to S. Dawes, DSIR, Auckland) in 1980. The natural range is from Venezuela to northern Argentina. Flowers are tristylous and seed set is rare.

References

Hodge, W.H. (1951) Three native tuber plants of the high Andes. Economic Botany 5(2), 185–201.

Hill, A.W. (1939) The oca and its varieties. Kew Bulletin (4), 169–173.

See also: Rooney, D. Country Diary. The Press 29/6/85

Anon (1986) Scientists groom Andean tubers for Ethiopia. New Scientist, no 1505, p 34.

Pachyrhizus erosus (L.) Urban

Leguminosae

Common

Yam Bean, Jicama (Mexico), Sinkamas (Phillipines), Yaka (Fiji).

Synonyms

Dolichos erosus L., *Pachyrhizus bulbosus* (L.) Kurz, *P. jicamas* Blanco.

Figure 36: Oca cultivars.

Notes
A tuberous rooted legume from Central America, now widely cultivated throughout the tropics and sub–tropics. The tubers are becoming increasingly popular in S.E. Asia and Fiji and some are being exported to United States markets mainly from Mexico and Hawaii. Mature seeds and foliage are toxic but the young pods are occasionally eaten. The closely related *P. tuberosus* (Lam.) Spreng. from South America produces larger tubers. The seed pods are covered with irritating hairs.

References
Kay, D.E. (1973) TP1 Crop and Product Digest No 2 Root Crops. Tropical Products Institute, London. pp 240–245.

Duke, J.A. (1981). Handbook of legumes of world economic importance. Plenum Press, New York. pp 182–184.

Panax quinquefolius L.

Araliaceae

Common
American ginseng.

Synonyms
Aralia quinquefolia Decne et Planch.

Notes
A deciduous perennial with a short rhizomatous stem at the apex of a thickened tap root. The dried root is the ginseng of commerce. A free draining soil and partial shade preferably from broadleaf forest are required for growing this crop. The root takes 4 to 5 years to obtain harvestable size after germination which itself may take 18 months. Trials were conducted at Lincoln in the late 1970's and are reported on in Palmer & Hurndell (1986).
N.B. Three other species of Panax are sold as ginseng:
P. ginseng C.A. Meyer, syn. *P. schinseng* Nees, Chinese or Korean ginseng.
P. notoginseng (Burk.) F.H. Cheng. Sanchi ginseng from southern China.
P. japonicus C.A. Meyer. Japanese ginseng. This is only used in Japanese medicine.

References
Palmer, J.A. and Hurndell, L.C. (1986). Ginseng. Crop Research Division, DSIR, Report no. 121. 15 pp.

Williams, L. and Duke, J.A. (1978). Growing ginseng. USDA Farmers Bulletin No. 2201, Washington. 8 pp.

Pandanus tectorius Solander

Pandanaceae

Common
Screw palm, Fara (Tahiti), Hala (Hawaii), Fala (Samoa), Hara or ara (Cook Is), Vadra or Balawa (Fiji), Pandon (Philippines).

Synonyms
Pandanus odoratissimus L. f., *P. odorifer* (Forsk.) O. Kuntze.

Notes
A small tree with spirally arranged leaves and prop–roots. Widely distributed in the Pacific and Australasia. The fruit were at one time an important food crop in Micronesia and parts of Polynesia. Several botanical varieties have been recorded. The male flower spikes are surrounded by fragrant edible bracts. A scent – 'kewda oil' is extracted from the flowers in India.

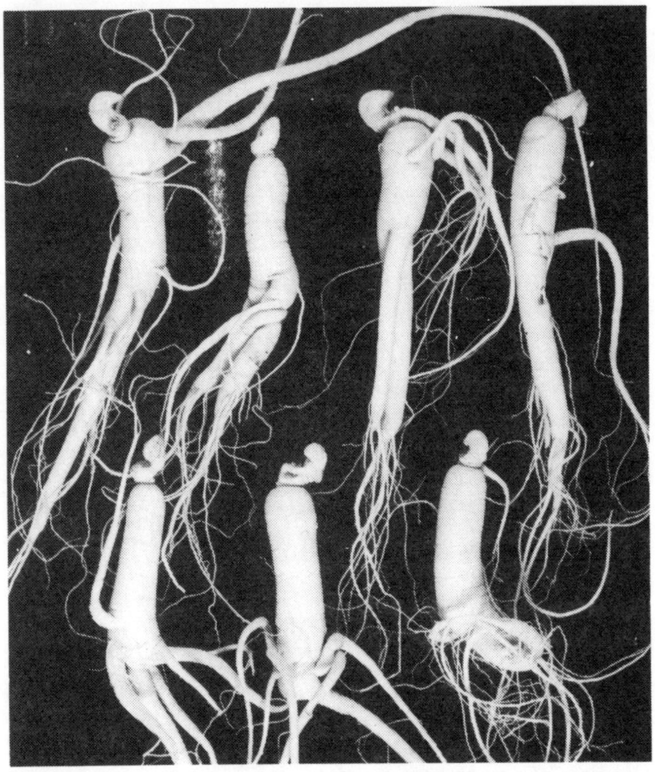

Figure 37: Roots of American ginseng. Note the short rhizome with apical bud.

Figure 38: Pandanus tectorius.

References

A good review of native use of Pandanus in the South Pacific is given in:

Brown, F.B.H. (1931). Flora of Southeastern Polynesia, 1, Monocotyledons. Bernice P. Bishop Museum, Bulletin 84, Honolulu. pp 30–37.

Miller, C.D. et al (1956). The use of Pandanus fruit as food in Micronesia. Pacific Science 10(1), 3–16.

Uhe, G. (1974). Wayside plants of the South Pacific. Stocton House, Auckland. pp 109–110.

See also: Poucher, W.A. (1979). Perfumes, cosmetics and soaps. 7th edition. Chapman and Hall, London. Vol. 1, pp 210–211 (kewda oil).

N.B. Other species of *Pandanus* are important food plants in New Guinea. Vide: Hyndman, D.C. (1984). Ethnobotany of Wopkaimin Pandanus: Significant Papua New Guinea plant resource. Economic Botany 38(3), 287–303.

The Malesian species with scented leaves often used in Southeast Asian cooking is *P. amaryllifolius* Roxb. Vide: Stone, B.C. (1978). Studies in Malesian Pandanaceae XVII. On the taxonomy of 'pandan wangi'. A pandanus with scented leaves. Economic Botany 32(3), 285–293.

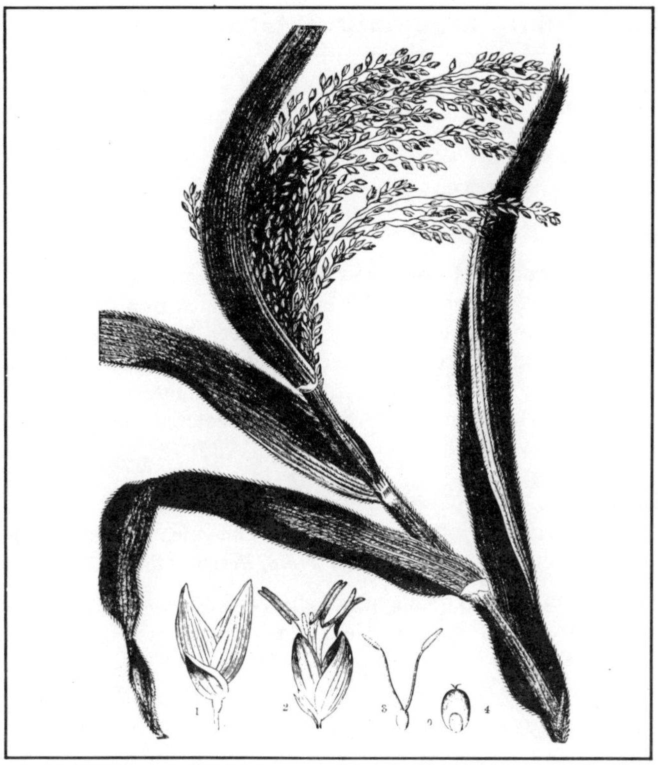

Figure 39: Common millet.

Panicum miliaceum L.

Gramineae

Common
Common millet, Proso millet, Hog millet, Broom–corn millet, Russian or French millet.

Notes
A very ancient crop species, not known in the wild. Cultivated in southern Europe since pre–historic times. Now grown in Eastern Asia, Central and Eastern USSR and parts of the Middle East. The crop is adapted to low water use on poor soils. The hulled grain is occasionally sold in N.Z. grocery stores as a 'health food'.

References
Purseglove, J.W. (1972). Tropical crops – Monocotyledons. Longman, London. Vol. 1, pp 199–200.

Mann, H.H. (1946). Millets in the Middle East. Empire Journal of Experimental Agriculture 14, 208–216.

N.B. *P. decompositum* R. Br. is the Australian 'Native millet'. *P. sumatrense* Roth. Little millet is cultivated to a limited extent in India and adjacent countries and may be substituted for common millet.

Papaver bracteatum Lindley

Papaveraceae

Common
Codeine poppy, Bracteate poppy, Scarlet poppy*.

Notes
A red flowered perennial poppy from Iran and adjacent countries and a diploid member of the ornamental oriental poppies (*P. orientale*) series. The major alkaloid constituent of the pods is 'thebaine' which is not a narcotic and can readily be converted to codeine. The bracteate poppy was grown in trials at Lincoln and in MacKenzie County from 1978 to 1983 and pods supplied to the Pharmacy School, University of Otago for the experimental production of codeine (vide Gopal 1984). Seed is now maintained in the DSIR Gene Bank at Lincoln.

References
Nyman, U. and Bruhm, J.G. (1979). *Papaver bracteatum* – a summary of current knowledge. Planta Medica 35(2), 97–117.

Levy, A. et al. (1981). Genetic improvement of *Papaver bracteatum.* Mortality and selection response of thebaine and seed yields. Planta Medica 43(1), 71–76.

See also: Gopal, Y.C. (1984). The production of codeine from a sample of thebaine rich *Papaver bracteatum.* A project report submitted as part of the course requirements for the degree of Bachelor of Pharmacy, University of Otago. 37 pp.

*Scarlet poppy is the preferred common name given in W. Hartley in 'Checklist of economic plants of Australia', CSIRO, Melbourne 1979.

Paracitrullus vulgaris Watt. *var. fistulosus*

Cucurbitaceae

Common
Tinda (Hindi), Squash melon.

Synonyms
Citrullus vulgaris var. fistulosus (Stocks) Duthie and Fuller.*

Figure 40: Bracteate poppy, Lincoln, November 1981.

Notes
Widely grown in northern India which is probably its region of origin. Occasionally grown by Indian migrants in the Pacific. Fruit is harvested at an immature stage whilst still hairy and the seeds are soft.
* N.B. *Citrullus lanatus* (Thunb.) Matsum. et Nakai (syn *C. vulgaris* Schrad.) is the common water melon. Water melons are mostly diploids with 11 pairs of chromosomes. *Paracitrullus* has 12 pairs.

References
Singh H, et al. (1982). Effect of different doses of nitrogen and phosphorus on growth, flowering and fruiting, seed yield and quality of tinda. Indian Journal of Horticultural Science 39(182), 94–100.

Anon (1950). The Wealth of India – Raw Materials, Vol 2. CSIR, New Delhi. p 188.

Parthenium argentatum Gray

Compositae

Common
Guayule (pronounced wy–oo–lee).

Notes
A desert shrub from Southwest Texas and Northern Mexico. It produces a rubber latex similar to that obtained from the rubber tree, *Hevea brasiliensis* and has long been a minor source of natural rubber. A project started to cultivate this plant during World War 2 was abandoned in 1946, but interest was rekindled in the U.S.A. in 1975. Some years later a trial plot was established at Lincoln College (Taylor, 1985).

References
National Academy of Sciences (1977) Guayule. An alternative source of natural rubber. NAS, Washington. 80 pp.

Taylor, J.O. (1985). Guayule – the New Zealand experience Journal of the Royal N.Z. Institute of Horticulture 13, 49–51.

Pelargonium graveolens L'Herit. ex Ait.

Geraniaceae

Common
Rose geranium.

Synonyms
Geranium radula Roth., *Pelargonium asperum* Erh. ex Willd. *P.intermedium* Knuth.
NB: *Pelargonium terebinthaceum* Cav. = *P. graveolens* cv. Little gem.

Notes
An erect, banched, shrubby perennial(to 1.3m) from South Africa, cultivated as a source of 'geranium oil' in France, Spain, Italy, Reunion, Kenya (where it is called 'Mowah') and North Africa. In New Zealand rose geraniums have escaped from gardens in north Auckland and have been recorded as adventives in the North Hokianga area and Waipoua State Forest (vide Sykes, 1982, N.Z. J. Bot. 20: 334). But there is still some question as to the true identity of some feral *Pelargonium* in N.Z. (Vide Sykes, 1988, Flora of New Zaeland, Vol.4, p738). Under natural conditions *P. graveolens* occurs in moist habitats in partial shade. Soil type and climate greatly affect the quality of geranium oil.
Other species producing geranium oil:
P. australe Willd. An Australian native. Mentioned by Buisson (1979). Potential crops for processing in N.Z. Pt 3.
P. capitatum (L.) L'Herit. occasionally wild in the Hokianga district of northern N.Z.
P. x fragrans Willd. (= *P. odoratissimum* x *P. exstipulatum*).
P. radens H.E. Moore (syn. *P. radula* L'Herit.)
P. x asperum Erh. ex Willd. (syn *P. roseum* Willd.). This hybrid has established itself at a number of sites on both islands of New Zealand (vide Sykes 1988, op. cit.).
Moore, H.E. (1955). Pelargoniums in cultivation. Baileya 3: 5–25, 41–46, and 70–97, gives the following list of fragrances:
Rose – *P. graveolens, P. capitatum* .
Lemon – *P. crispum* .
Peppermint – *P. tomentosum* .
Pine – *P. denticulatum* .
Apple – *P. odoratissimum*
Apricot – P. 'M. Ninon'.
Coconut – *P. grossularioides* .
Nutmeg – *P. x fragrans* . Some plants of this hybrid are established on volcanic cliff faces in the Port Hills near Christchurch.
Pungent – *P. quercifolium, P. radens* .
NB: Strictly speaking true 'geranium oil' is obtained from *Geranium macrorrhizum* L., native to Bulgaria.

References
Walt, J.J.A.van der, and Vorster P.J.(1988). Pelargoniums of Southern Africa, Volume 3. National Botanic Gardens, Kirstenbosch. pp 63–65.

Poucher, W.A. (1975). Perfumes, cosmetics and soaps. 7th edition. Chapman and Hall, London. Vol. 1, pp 173–175.

Lawrence, B.M. et al (1975). The chemical composition of some Pelargonium species. International Flavours and Food Additives 6(1), 42.

See also: Guenther, E. (1952). Recent developments in essential oil production. Economic Botany 6(4), 355–378.

Figure 41: Pearl millet.

Pennisetum glaucum (L.) R. Br.

Gramineae

Common
Pearl millet, Bulrush millet, Spiked millet, Cat–tail millet, Bajra (India).

Synonyms
Pennisetum typhoides (Burm.) Stapf and C.E. Hubbard, *P. typhoideum* (L). C. Rich., *P. americanum* (L.) Leake.

Notes
A small grain crop adapted to dry sandy areas of the tropics and sub–tropics. It is important in the Sahel zone of West Africa and some parts of India. There are a number of botanical races of this species.

References
Rachie, K.O. and Majmudar, J.V. (1980). Pearl millet. Pennsylvania State University Press, University Park. 307 p.

Purseglove, J.W. (1972). Tropical crops – Monocotyledons. Longman, London. Vol. 1, pp 202–214.

Perilla frutescens (L.) Britt.

Labiatae

Common
Perilla, Beef steak plant, Shiso (Japanese for var. crispa), Egoma (Japanese for var. Japonica).

Synonyms
Perilla nankinensis Decne, *P. ocimoides* L.

Notes
An annual herb from eastern Asia. There are three principal botanical varieties:
1. *P. frutescens var. crispa* Decne. with both purple and green leaved cultivars, is used as a flavouring herb in China and Japan. Young seedlings (sprouts) are eaten raw, boiled or pickled. Leaves and flowers are often salted and used as a condiment or fried in vegetable oil. Young plants are eaten raw with fish in Japan and leaves and stems of the purple cultivar are used to colour pickled fruit, Japanese artichokes (*Stachys affinis*) and ginger. Seed of var. *crispa* has been available for some time in New Zealand but the plant is little used here.
2. *P. frutescens var. acuta* Kudo. The dried leaves of this variety are used in medicine (Koezuka, 1985 below) and are included in the Pharmacopoeia of Japan (1982).
3. *P. frutescens var . japonica* (Hassk.) Hora., is grown as an oil seed crop from Japan to northern India. The oil is edible but most often used for industrial purposes.

References
Yashiroda, K. (1968). A dictionary of popular Japanese herbs and their uses today. Plants and Gardens 24(2), 44–58. Reprinted as Brooklyn Botanic Garden, Handbook 57.

Burkill, I.H. (1935). Economic products of the Malay Peninsula. The Crown Agents, London. Vol. 2, p 1694.

See also: Ishikura, N. (1981). Anthocyanins and flavones in leaves and seeds of Perilla plant. Agricultural and Biological Chemistry 45(8), 1855–1860.

Koezuka, Y. et al. (1985). An intestinal propulsion promoting substance from *Perilla frutescens* and its mechanism of action. Planta Medica 6, 480–482.

Figure 42: Curled bronze leafed perilla.

Petasites japonicus (Sieb. et Zucc.) Maxim.

Compositae

Common
Sweet coltsfoot, Butterbur, Fuki (Japan), Akitabuki (Japanese for var. giganteus)

Synonyms
Nardosmia japonica Sieb. et Zucc.

Notes
A deciduous herb with large orbicular leaves (to 30 cm diam.) on petioles up to 60 cm long. The flowers are dioecious and appear before the leaves in late winter. *P. japonicus var giganteus* (F. Schmidt) Nichols from northern Japan is appreciably larger with petioles up to 2 m long and leaves more than 1 m in diameter. Only the common form has been introduced to New Zealand and all the plants observed by the writer have been female. The young leaf stalks and flowering shoots are used as a vegetable in Japan, Korea, northern China and the Ryukyu Islands.

References
Yashiroda, K. (1968). A dictionary of popular Japanese herbs and their uses today. Plants and Gardens 24(2), 44–58.

Figure 43: Sweet coltsfoot. The ruler is 40 cm long.

Ohyama, I. (1968). Akita–buki, giant sweet coltsfoot. Plants and Gardens 24(2), 25–26.

Phaseolus acutifolius A. Gray

Leguminosae

Common
Tepary bean.

Notes
An indeterminate twining annual native to arid areas from Arizona south to Guatemala. They require more moisture for germination and growth than true desert plants but are a crop well adapted to hot arid conditions. In cultivars the beans are small, frequently white but also occur in many colours. Attempts to promote tepary cultivation in the arid southwest of the United States earlier this century failed mostly due to public dislike of the often unattractive, strong flavoured beans. Teparies are now only produced on a small scale in Mexico.

References
Nabham, G.P. and Felger, F.S. (1978). Teparies in southwestern North America. Economic Botany 32(1), 2–19.

National Academy of Sciences (1979). Tropical legumes: Resources for the future. NAS, Washington. pp 92–97.

N.B. The major edible species of the genus *Phaseolus* show a range of climatic adaptions from cool moist to arid warm in the following order: *P. coccineus* , the scarlet runner; *P. vulgaris* , common beans; *P. lunatus* , lima beans; *P. acutifolus* , tepary beans.

Phaseolus lunatus L.

Leguminosae

Common
Lima bean, Sieva bean, Patani (Philippines), Kachang China (Malaysia), Aoi–mame (Japan).

Synonyms
Phaseolus limensis Macf.

Notes
A twining annual plant with relatively short broad pods containing few seeds. Semi determinate (bush) cultivars have been selected. Pods are invariably fibrous. Most cultivars are white seeded but seeds of all colours occur. The characteristic flavour is due to the presence of prussic acid. In some old tropical cultivars this could occur in toxic quantities (Vide Vega and Sotelo, 1986). Lima beans require higher temperatures for growth than common beans (*P. vulgaris*) but are more drought tolerant. Large seeded cultivars tend to be less cold tolerant than those with smaller seeds. Lima beans have not been known to crop successfully in New Zealand and are rarely planted in the islands of the South Pacific.

References
Vega, A. de la and Sotelo, A (1986). The nutritional quality and toxin content of wild and cultivated lima beans. Qualitas Plantarum 36(1), 75–83.

Fisher, V.J. and Weaver, C.K. (1974). Flowering, pod set and pod retention of lima beans in response to night temperature, humidity and soil moisture. Journal of the American Society of Horticultural Science 99(5), 448–450.

Phormium tenax Forst. et Forst. f.

Agavaceae

Common
New Zealand flax, N.Z. hemp, Harakeke (Maori).

Notes
An important fibre source for the Maori which became a major 19th century New Zealand export. Commercial production declined early in the 20th century due to competition from tropical fibres. A living collection of Maori cultivars is held at DSIR Centres at Lincoln and Havelock North.

References
Critchfield, H.J. (1951). *Phormium tenax:* New Zealand's native hard fibre. Economic Botany 5, 172–184.

Brooker, S.G. et. al (1988). Economic native plants of New Zealand. Botany Division, DSIR, Christchurch. pp 39–40.

See also: Boyce W.R. (1947). Phormium research. Notes on a talk on 16/6/47. Bulletin of the Wellington Botanical Society. No. 17, 2–4.

Griffin, G.W. (1884). New Zealand, her commerce and resources. Government Printer, Wellington. pp 128–132.

Report of Paralimentary Committee (1890). The flax industry. The N.Z. Country Journal 14(5), 415–417.

Phyllanthus acidus (L.) Skeels

Euphorbiaceae

Common
Otaheite gooseberry, Star gooseberry, Iba (Phillipines), Chermai (Malay), Kavakava (Rarotonga), Vini (Samoa).

Synonyms
Cicca acida Merr.

Notes
A small tree with acid fruit. Widespread in Pacific region but unimportant and mostly confined to home gardens.

References
Page P.E. (1984) Tropical tree fruits for Australia. Queensland Department of Primary Industries, Brisbane. p 61.

Allen, Betty. M. (1967) Malayan fruits. Donald Moore Press, Singapore. pp 45–46.

Phyllanthus emblica L.

Euphorbiaceae

Common
Emblic myrobalan, Indian gooseberry, Melaka (Malay), Aonla (Hindi), Nelli (Tamil).

Synonyms
Emblica officinalis Gaertn.

Notes
A wide crowned deciduous fruit tree from lowland forest areas of Southeast Asia. The astringent acid tasting fruit are bourne at the ends of leafy twigs. They are mainly used for pickles and chutney. Leaves, bark and young fruit are used in dyeing and tanning. The dried fruit is a common constituent of Eastern medicine and is official in the Chinese Pharmacopoeia (1977). Emblic is grown as an ornamental tree in Hawaii but is rare or absent from most other Pacfic Islands except Fiji.

References
Singh, I.S. and Pathak, R.K. (1987). Evaluation of aonla varieties for processing. Acta Horticulturae 208, 173–175.

Morton, Julia F. (1960). The emblic *(Phyllanthus emblica).* Economic Botany 14(2), 119–128.

Phyllostachys heterocycla (Carr.) Mitf.

Gramineae

Common
Moso bamboo (U.S.A.), Moso–chiku (Japan). NB. Numerous cultivars have been recorded including the tortoise–shell bamboo, Kikko–chiku in Japanese.

Synonyms
Bambusa heterocycla Carr., *Phyllostachys pubescens* Mazel. ex Houz. de Lehaie, *P. edulis* Houz. de Lehaie.

Notes
A large spreading bamboo from Eastern China, probably introduced to Japan where it is widely cultivated. The culms may be up to 20 m tall and 20 cm in diameter under favourable situations but are generally smaller in New Zealand, especially south of Auckland. It is hardy as far south as Christchurch in sheltered sites. At least two cultivars are known in N.Z.: *P. heterocycla cv kikko–chiku,* Mitford ex Ohwi, the tortoise–shell bamboo and *P. heterocycla f. pubescens* (Houz. de Lehaie) Muroi. Vide: D. Ohrnberger and J. Goerrings (1983). The bamboos of the world, genus *Phyllostachys* . Odenthal, West Germany.

References

Porterfield, W.M. (1938). Bamboo as it is grown for food in China. Journal of the New York Botanic Gardens 39, 282–288.

Palmer, J. (1981). Bamboo shoots. Tree Crops Journal 6(3), 33–35.

See also: Gerlach, J.C. (1974). The planting and care of bamboos MAF, Ruakura Agricultural Research Centre. 9 pp.

N.B. F.A. McClure (1957). Bamboos of the genus Phyllostachys under cultivation in the United States. US Department of Agriculture Handbook 114 Washington, lists 18 edible species of this genus. Those included in the collection held at John Isaachsen's Bamboo nursery, West Coast Road, Oratia are (using Ohrnberger and Goerring's nomenclature with the common N.Z. name in parentheses):

(angusta) *P. angusta* McClure. Stone bamboo.

(aurea) *P. aurea* Carr. The golden bamboo, Fishpole bamboo (US).
N.B. This species was at one time imported from Japan for use as fishing poles. However 'split cane' rods are usually made from Tonkin cane, Hotei–chiku (Japan). *Arundinaria amabilis*, a species which appears to have died out in New Zealand.

(bambusoides) *P. bambusoides* Sieb. et Zucc. The giant timber bamboo or 'madake' of Japan. Numerous cultivars exist, including:

(violascens) *P. bambusoides var. violascens* Crouzet.

(marliacea) *P. bambusoides f. marliacea* (Mitford) Muroi.

(sulphurea) *P. bambusoides f. holochrysa* Pfitzen.

(nigra) *P. nigra* (Loddiges) Munro. Black bamboo, Kuro–chiku (Japan).

(boryana) *P. nigra f. boryana* (Mitford) Muroi.

(mitis) *P. nigra f. henonis* (Bean) Muroi.

(puberula) *P. nigra f. punctata* (Bean) Schelle.

(vivax) *P. vivax*. McClure.

Other Phyllostachys species known to be in New Zealand include: *P. makinoi* Hayata and *P. meyeri* McClure.

Edible bamboos in other genera:

1. Philippine Is.
    - *Bambusa blumeana* Schultes – Kauayan tinik.
    - *B. vulgaris* Schrad. – Kauayan kiling.
    - *Dendrocalamus merrillianus* Elm. – Bayog.
    - *D. latiflorus* Munro. – Machiku bamboo.

Figure 44: Phyllostachys heterocycla.

    - *D. asper* (Schult.) Backer – Giant bolo.
    - *Gigantochloa laevis* (Blanco) Merr. – Bolo.
2. Taiwan.
    - *Leleba edulis* Odashima. Ho–ryokutiku (Japanese)
    - *Dendrocalamus latiflorus* Munro. Machiku.
    - *Phyllostachys heterocycla* (Carr.) Mitf. Only minor usage.
3. Eastern China and Japan.
    - *Phyllostachys heterocycla* (Carr.) Mitf.
    - *P. bambusoides* Sieb. & Zucc.
    - *P. viridis* McClure
    - *P. dulcis* McClure

In southern China, *Bambusa vulgaris* is used. But this species is too tropical to survive unprotected in New Zealand. There is a specimen in the Begonia House of Wellington Botanic Gardens.

Physalis ixocarpa Brot.

Solanaceae

Common
Mexican or Mayan husk tomato, Tomate mexicano, Tomatillo, Jam berry.

Synonyms
Physalis aequata Jacq.

Notes
An important vegetable of the Aztec and Mayan civilisations. Cultivation of the tamatillo has persisted in areas of Mexico and Guatemala. It has been introduced to South Africa for processing and to East Africa where it has become a weed. In New Zealand a small fruited cultivar was found in gardens of South Auckland in the late 1960's. A further introduction of larger fruited but less flavoured cultivars was made by Horticultural Processing Division about 1980. The fruit of this species is really only suitable for processing.

References
Quiros , C.F. (1984). Overview of the genetics and breeding of husk tomato. Hort Science 19(6), 872–874.

Morton, Julia F. (1987). Fruits of warm climates. Creative Resource Systems, Winterville, NC. pp 434–437.

N.B. The cape gooseberry, *Physalis peruviana* L., is much better known in New Zealand and has been used commercially for jam making in Australia.

Vide: Klinac, M.L. and Wood, F.H. (1986). Cape gooseberry (*Physalis peruviana*). The Orchardist of New Zealand 59(3), 103.

Figure 45: Physalis ixocarpa (right), with the North American ground cherry, P. pruinosa (left).

Pimenta racemosa (Mill.) J.W. Moore

Myrtaceae

Common
Bay rum tree, Bay oil tree, Pimenta (Samoa).

Synonyms
Caryophyllus racemosus Mill., *Pimenta acris* Kostel.

Notes
A tree (to 12m), originally from the West Indies, cultivated and occasionally naturalised in Fiji, Samoa, Cook and Society Islands. A fragrant oil distilled from the leaves and shoots is used in cosmetics and medicine. The dried green fruit of the related West Indian tree *P. dioica* (L.) Merrill is the 'allspice' or 'pimento' of commerce.

References

Purseglove, J.W. (1968). Tropical crops, Dicotyledons. Longman, London. p 409.

The Wealth of India, Raw materials, Vol 8. CSIR, New Delhi, 1969. pp 59–60.

Pinus pinea L.

Pinaceae

Common
Stone pine.

Notes
A largely coastal tree from southern Europe and Asia Minor. Cultivated for its edible seed known as 'pignolias' or 'pignoli' in Great Britain. The principal areas of production are in Spain and Italy. Pine cones are collected in summer with long hooked poles and spread in the sun for the seed to fall out. This species is widespread, but unused, in southern New Zealand. The arolla pine, *P. cembra* L. of alpine Europe above 2 000 m is also harvested in Switzerland and Austria and the related dwarf Siberian pine, *P. pumila* (Pall.) Reg. in Russia. In North America the principal species with edible nuts is *P. cembroides*. This species is grown but not utilised in New Zealand.

References

Trap, L. (1987). *Pinus pinea* : a tree of many uses. Growing Today 4(3), 13.

Bull, P., Jackson, D. and Bedford, T. (1985). Edible tree nuts in New Zealand. Government Printer, Wellington. pp 64–65.

Harrison, S.G. (1951). Edible pine kernels. Kew Bulletin Pt 3, 371–375.

Bean, W.J. (1976). Trees and shrubs hardy in the British Isles. 8th edition. John Murray, London, Vol. 3. pp. 237–238.

N.B. Several North American pines produce edible seed which were an important food source for the original inhabitants but are now marginally exploited. Vide:

Lanner, R.N. (1981). The pinon pine. A natural and cultural history. Nevada Press, Reno. 208 pp.

Fogg, G.G. (1966). The pinyon pines and man. Economic Botany 20(1), 103–105.

Egolf, R.L. (1977). The pine family (pinyons) and their relatives in E.A. Menniger (ed.). Edible nuts of the world. Horticultural Books, Stuart, Fl. pp 153–160.

Piper betle L.

Piperaceae

Common
Betel, Betel vine, Pan (Hindi).

Notes
A woody vine with dimorphic branching. The leaves are chewed together with betel nut in many parts of Southeast Asia. Cultivated to a limited extent by the Indian population of Fiji.

References

Ram–Jan, S. (1937). Cultivation of betel leaf in Fiji. Agricultural Journal, Fiji 8(4), 49.

Purseglove, J.W. (1968). Tropical Crops, Dicotyledons. Longman, London. pp 437–440.

Piper methysticum Forst. f.

Piperaceae

Common
Kava or ava (Polynesia), Yaqona (Fiji – pronounced yangona).

Synonyms
Macropiper methysticum Hook.

Notes
A shrubby herbaceous plant cultivated for its roots from which a traditional beverage is prepared in most Pacific Islands from New Guinea eastwards. The medicinal value of the roots of this herb are discussed in the British Pharmaceutical Codex, 1934 (p 573) and it is still included in the British Herbal Pharmacopoeia, 1983 (pp 162–3) for infections of the genito–urinary tract.

References

Gatty, R. (1956). Kava – Polynesian beverage shrub. Economc Botany 10, 241–249.

Steinmetz, E.F. (1960). *Piper methysticum* (kava) famous drug plant of the South Sea Islands. E.F. Steinmetz, Amsterdam. 46 p. reprinted as: Kava–kava, famous drug plant of the South Sea Islands. High Times/Level Press, New York, 1973.

See also: Safer, M. (1985). Yaqona and peripheral economy. Pacific Viewpoint 26(9), 415–436.

Piper nigrum L.

Piperaceae

Common
Pepper.

Notes
A perennial woody vine native to the Western Ghats of India. Introduced to Fiji before 1880 and occasionally grown elsewhere in the South Pacific. However pepper requires a humid tropical climate and few Pacific Islands would be considered as ideal commercial sites. The berries (technically a drupe), when dried are the black pepper of commerce. White pepper is prepared by retting fully ripe berries in water to remove the skin before drying.
Two other species are of local importance, especially in medicine, in Southeast Asia:
P. longum L. Long pepper. A slender climber from India. Whole spikes are harvested immature and dried.
P. cubeba L.f. Cubebs or tailed pepper from Indonesia.

References
Kay, D.E. (1970). The production and marketing of pepper (a review). Tropical Science 12, 201.

Woord P.W.F.de (1986). Current state and prospective trends of black pepper (*Piper nigrum* L.). Outlook on Agriculture 15(4), 186-195.

Pistacia vera L.

Anacardiaceae

Common
Pistachio nut.

Notes
Slow growing, drought resistant, deciduous, dioecious nut trees. Probably native to Central Asia but grown in the Middle East since pre–historic times. Cultivated in the U.S.A. and in South–eastern Australia since World War 2 and more recently introduced to New Zealand.

References
Maggs, D.H. (1982). An introduction of pistachio growing in Australia. CSIRO, Canberra (?). 36 pp.

Rosengarten, F. (1984). The book of edible nuts. Walker & Co., New York. pp 196–213.

See also: Block, F. and Brekke, J.E. (1960). Processing of pistachio nuts. Economic Botany 14 (2), 129–144.

Plantago ovata Forsk.

Plantaginaceae

Common
Ispaghula, Isabgol (Hindi).

Synonyms
Plantago isphaghula Roxb.

Notes
An erect annual plant native to the Mediterranean area and western Asia. Introduced to India at the time of the Mahomedan invasions from the north and now cultivated there. The mucilagenous seed husk from this species is a major medicinal product. There have been recent attempts to cultivate this species in Israel and in Western Australia on the Ord river near Konuwara with some success. Trials at Lincoln from 1981 to 1984 were a failure. Plants grew to little over 20 cms in height (as compared with an average of 40 cms at Konuwara) and were too short for mechanical harvesting.

References
Kalyanasundarum, N.K. et. al (1984). Psyllium, a monopoly of Gujerat. Indian Horticulture 28(4), 35–37 and 41.

Gupta, R. (1982). Recent advances in the cultivation of isabgol (*Plantago ovata* Forsk.) in India. In Atal and Kapur (eds.) Cultivation and utilisation of medicinal plants CSIR, Jammu–Tawi. pp 406–417.

Plantago psyllium L.

Plantaginaceae

Common
Psyllium, Fleawort.

Synonyms
Plantago arenaria Waldst. et Kit., *P. indica* L.

Notes
An erect annual, native to dry sites in southern, central and eastern Europe. It is cultivated principally in France and Spain for the mucilagenous seed husk which is used medicinally (psyllium husk). Trials were conducted at Lincoln between 1981 and 1985 and yields of up to 900 kg seed per ha. obtained. One commercial crop was grown in 1984/5 but failed to compete with imported seed of *P. ovata* from India.

Figure 46: Ispaghula plant.

Figure 47: Psyllium plant.

References

International Trade Commission (1982). Markets for selected medicinal plants. GATT, Rome. pp 167–171.

Morton, Julia F. (1977). Major medicinal plants. Charles C. Thomas, Springfield. pp 331–332.

UNESCO (1960). Medicinal plants of the arid zone, Paris. pp 47–48.

Podophyllum peltatum L.

Podophyllaceae

Common

May apple, Podophyllum, American mandrake, Wild jalap.

Synonyms

Podophyllum callicarpum Raf.

Notes

A deciduous herb native to damp deciduous woodland of North America from southeastern Canada south to Georgia. The dried root was once used medicinally as a powerful cathartic but recent interest has centred on the ability of the liquid extract in alcohol to cure genital warts. An alternative species from the Himalayan region, *P. emodi* Wall. has been grown in New Zealand gardens as an ornamental.

References

Meijer, W. (1974). *Podophyllum peltatum* – may apple, a potential new cash crop. Economic Botany 28(1), 68–70.

Morton, Julia F. (1977). Major medicinal plants. Charles C. Thomas, Springfield. pp 87–89.

Pogostemon cablin (Blanco) Benth.

Labiatae

Common
Patchouli, Kablin (Philippines), Patiole (Polynesia).

Synonyms
Pogostemon patchouli Hook.

Notes
An erect shrubby aromatic plant up to 1 m high, native to the Philippines but widely planted in South Asia and the Pacific. Cultivated particularly in Sumatra, Madagascar, the Seychelles and India for the aromatic oil extracted from dried foliage. In the Pacific Islands patchouli is widespread but only used spasmodically by individuals, mainly for washing hair. Patchouli oil is one of the best fixatives for heavy perfumes and imparts a characteristic odour to Indian shawls.

References
Soepadyo, R. and Tan, H.T. (1968). Patchouli; a profitable catch crop. World Crops 20(1), 48–52.

Lionnet, J.F.G. (1962). Seychelles patchouli. World Crops 14(10), 336–338 and 356.

See also: Poucher, W.A. (1979). Perfumes, cosmetics and soaps. 7th edition, Chapman and Hall, London. Vol. 1, pp 294–5.

Polymnia sonchifolia Poepp. et Endl.

Compositae

Common
Yacon, Jiquima.

Synonyms
Polymnia edulis Wedd.

Notes
A South American tuber of little economic value. Introduced into N.Z. c. 1980 as a private venture by an Auckland nurseryman. Tested in Italy prior to World War II as a possible source of sugar or forage.

References
Endt, A. (1983) The yacon and aniu, new from South America. Commercial Horticulture, July 1983, pp 14, 15 and 32.

Vietmeyer, N. (1984) The lost crops of the Incas. Ceres 17(3), 37–39.

Figure 48: Yacon plant in a Christchurch garden.

Pometia pinnata Forst. f.

Sapindaceae

Common
Langsir, Polynesian lychee, Taun (New Guinea), Dawa (Fiji), Tava (Samoa and Tonga), Kava (Tahiti).

Synonyms
Nephelium pinnatum Cambess.

Notes
A large lowland forest tree. Flowers in terminal panicles. Fruit, often dull red coloured and up to 7 cm in diameter, are edible. The oily seeds are also occasionally roasted and eaten. A very variable species (Jacobs, 1962) occuring naturally through western Polynesia (but not the Cook Islands), Papua New Guinea, west to Sri Lanka and north to Taiwan. Several cultivars are reported in Fiji (Parham, The plants of Fiji. Government Printer, Suva, 1962). This species would appear to be very little known and is under–utilised.

References
Watson, B.J. (1984). Taun or dawa. In P.E. Page (ed). Tropical tree fruits for Australia. Queensland Department of Primary Industry, Brisbane. pp 204–206.

Whitman, W.F. (1972). The olosapo, the sunsapote and the Fijian longan. Proceedings of the Florida State Horticultural Society 84, 323–325.

See also: Jacobs, M. (1962). *Pometia (Sapindaceae)* . A study of variability. Reinwardtia 6(2), 109–144.

N.B. *Pometia pinnata* leaves are important in New Guinea yam gardens Vide: Allen, B.J. (1982). Yam gardens and fallows in the Torricelli foothills. In: Bourke and Kesavan (eds). Proceedings of the Second Papua New Guinea Food Crops Conference, Port Moresby. pp 236–239.

Portulaca oleracea L.

Portulacaceae

Common
Purslane, Pigweed (feral forms), Taukuku ni vuaka (Fiji), Tamole (Samoa and Tonga), Pokea (Rarotonga), Aturi (Tahiti).

Synonyms
Portulaca laevis Wall., *P. officinarum* Crantz, *P.olitoria* Pallas, *P. sativa* Hew., *P. suffruticosa* Thw.

Notes
A prostrate succulent annual herb. A cosmopolitan weed of warm temperate and tropical areas including much of lowland New Zealand. May have been an early introduction into the South Pacific. Feral forms are edible but now seldom utilised. European cultivars are larger, more leafy polyploids. The flowers in some cultivars tend to be cleistogamous. Purslane makes an attractive summer salad crop for hot dry sites, but water stress tends to initiate flowering. Two related species, *P. lutea* Soland. ex Forst. and *P. johnii* Poelln. are used as food plants by the atoll dwellers of the central Pacific (Turbott, 1954).

Figure 49: Purslane cultivar in December (The brick c. 11cm wide).

References
Herklots, G.A.C. (1972). Vegetables in Southeast Asia. G. Allen and Unwin, London. pp 177–179.

Gorman, Marion.(1988). Purslane goes American. Organic Gardening 35(6), 68–69.

Danin, A. et al. (1978). Cytogeography and taxonomy of Portulaca polyploid complex. Israel Journal of Botany 27, 177–211.

See also: Turbott, I.G. (1954). Portulaca a speciality in the diet of Gilbertese in the Phoenix Islands, central Pacific. Journal of the Polynesian Society 63(1), 77–85.

Psidium guajava L.

Myrtaceae

Common
'Guava', with minor derivatives, is almost universal.

Notes
A small to medium sized evergreen tree from Central America. Now introduced to sub–tropical areas world–wide. Very few trees occur in northern New Zealand but guavas are an important weed species in Fiji and some other Pacific Islands. N.B. The hardier cattley guava (*P. cattleianum* Sabine.) is relatively common in New Zealand gardens. Both red and yellow fruited forms occur but are little utilised. Other species of *Psidium* are cultivated in South America (vide: Morton J.M. (1987). Fruits of warm climates. Creative Resource Systems, Winterville. pp 363–367.)

References
Shiguera, G.T. and Bullock, R.M. (1983). Guava *(Psidium guajava)* in Hawaii, – history and production. University of Hawaii, Research Extention Series 035. 28 pp.

Ruehle, G.D. (1948). The common guava – a neglected fruit with a promising future. Economic Botany 2, 306–325.

Batten, D.J. (1984). Guava, *Psidium guajava* L.. In P.E. Page (ed.), Tropical tree fruits for Austalia. Queensland Department of Primary Industries, Brisbane. pp 113–120.

Figure 50: Unicorn plant.

Proboscidea louisianica (Mill.) Thell.

Martyniaceae

Common
Unicorn plant, Ram's horn, Devil's claw.

Synonyms
Proboscidea jussieui Steud.

Notes
A herbaceous spreading plant native to the southern United States grown as an ornamental and for the young fruit which may be pickled in the same manner as young cucumbers. The fruit matures as a woody capsule with a long, curved beak beneath a thick fleshy deciduous skin. As the capsule dries the beak splits into 2 curved hooks. At this stage they are frequently used in dried flower arrangements. The unicorn plant is a weed in parts of Australia but in New Zealand it is only grown as an ornamental pot plant.

References
Lawrence, G.H.M. (1957). *Proboscidea* and other unicorn plants. Baileya 5, 127–132.

Whittet, J.N. (1958). Weeds. The Government Printer, Sydney. pp 321–322 (as purple flowered devil's claw).

Psophocarpus tetragonolobus (L.) DC.

Leguminosae

Common
Winged bean, Goa bean, Seguidillas (Spanish – Philippines), Kachang botol (Malay), Bin (Papua New Guinea – Pidgin).

Synonyms
Dolichos tetragonolobus Stikm.

Notes
A climbing perennial legume with edible pods and roots. Recently this species has been extensively promoted because of high protein content of the legumes (pods). However, in its native areas of southeast Asia and New Guinea these have never been greatly appreciated. They are flavourless and rapidly become fibrous. In New Guinea some cultivars are grown for their starchy roots but even these are a poor substitute for *Pachyrhizus*. Young shoots parasitised by the fungus, *Woroninella psophocarpa,* which causes them to become swollen and orange coloured are regarded as a delicacy in parts of Indonesia.

References
National Academy of Science (1975). The winged bean. A high protein crop for the tropics. National Academy of Science, Washington. 43 pp.

Hymowitz, T. and Boyd, J. (1977). Origin, ethnobotany and agricultural potential of the winged bean. Economic Botany 31(2), 180–188.

Pteridium esculentum (Forst. f.) Cockayne

Dennstaedtiaceae

Common
Bracken, Aruhe (N.Z.).

Synonyms
Pteridium aquilinum (L.) Kuhn, the European bracken, is now regarded as a separate species.

Notes
A great deal of botanical confusion has occurred with this species (vide Hatch, 1984). The starchy rhizomes of bracken were at one time a staple food of the Maori in New Zealand, particularly in the southern regions. This is discussed by Crowe (1981). The young shoots – fiddle heads, are eaten by the Japanese as they were at one time in Europe. However, poisoning of cattle by bracken has long been recorded and the toxic principle shown to be transmitted through the milk. More recently the presence of an active carcinogen, shikimic acid, has been studied (Evans, 1976) and found to be particularly concentrated in the young shoots.

References
Hatch, E.D. (1984). George Forster and the tough morsel. Journal of the Auckland Botanical Society 40(1), 15–17.

Crowe, A. (1981). A field guide to the native edible plants of New Zealand. Collins, Auckland. pp 114–116.

See also: Evans, I.A. (1976). Relationship between bracken and cancer. Botanical Journal of the Linnean Society 73, 105–112.

N.B. Other fern species are occasionally eaten in the Malay–Asia region vide:

Copeland, E.B. (1942). Edible ferns. American Fern Journal 32(4), 121–126.

Hodge, W.H. (1973). Fern foods of Japan and the problem of toxicity. American Fern Journal 63(3), 77–80.

and elsewhere:

May, L.W. (1978). The economic uses and associated folklore of ferns and fern allies. The Botanical Review 44(4), 491–528.

Pueraria lobata (Willd.) Ohwi

Leguminosae

Common
Kudzu, Kuzu (Japan), Aka (Tonga), Yaka (Fiji).
NB 'Yaka' also refers to the yam bean *(Pachyrhizus erosus)*. There are many names for this species in New Caledonia and New Guinea.

Synonyms
Dolichos triloba L., *D. lobatus* Willd., *Pueraria triloba* Makino, *Pueraria thunbergiana,* Benth; *Pachyrhizus trilobus* DC.

Notes
A strong growing scandent legume with a large starchy root. The root is harvested for both food and medicine in the Far East and was at one time an important food source in Melanesia and New Caledonia. Current usage is restricted by the bland taste and large fibre content. Widely distributed through warm temperate and sub–tropical regions as a forage crop and for erosion control. Although kudzu flowers regularly in Fiji it does not seed and no flowering has been recorded in limited trials in New Zealand.

References
Tanner, R.D. et al. (1979). Kudzu: Potential agricultural and industrial resource. Economic Botany 33, 400–412.

Herklots, G.A.C. (1972) Vegetables in Southeast Asia. George Allen and Unwin, London. pp 468–470.

Porterfield, W.M. (1938). Ko, the kudzu vine, provides food, shade, clothes and medicine. Journal of the New York Botanical Gardens 39, 203–205.

See also:– Barrau, J. (1965) Witness of the past: Notes on some food plants of Oceania. Ethnology 4, 262–294.

Watson, J.B. (1968). Pueraria: names and traditions of a lesser crop in the Central Highlands, New Guinea. Ethnology 7, 268–279.

Punica granatum L.

Punicaceae

Common
Pomegranate, Limoni (Samoa), Pamikanite (Tonga).

Notes
A small, long–lived, tree, more or less spiny. Native to the Middle East but cultivated throughout warm temperate and subtropical areas with a semi–arid climate. Pomegranates are adapted to regions with cool winters and hot summers. Under cooler conditions trees frequently do not fruit. The pomegranate is little known in New Zealand or in the islands of the South Pacific. Trees grow well in the drier areas of Fiji, but are rare.

References
Morton, J.F. (1987). Fruits of warm climates. Creative Resource Systems Inc., Winterville, N.C. pp 352–355.

Goor, A. (1967). The history of the pomegranate in the Holy Land. Economic Botany 21(2), 215–229.

Creasy, R. (1986). The gardener's handbook of edible plants. Sierra Club Books, San Francisco. pp 235–238.

Quercus suber L.

Fagaceae

Common
Cork oak.

Notes
A variable species native to the western Mediterranean and cultivated principally in Spain and Portugal. Trees do not yield a crop of cork until they are around 20 years old, and do not reach optimum production until twice this age. The first cut is very rough and of little value. Subsequent harvesting is at intervals of 9 or 10 years. For these reasons commercial production outside the natural range of this species has seldom been attempted. The cork oak was introduced to New Zealand in 1855 (Gillies, 1883) and trees are widely scattered over both islands.

References
Long, O. (1985). The cork oak in New Zealand. Growing Today 2(5), 41–42.

Mentor, J.R. (1986). Cork oaks as a hobby. Growing Today 3(4), 29.

Stockwell, P. (1947). The culture of cork oak in Spain. Economic Botany 1(4), 381–388.

Cooke, G.B. (1948). Cork and cork products. Economic Botany 2(4), 393–402.

Gillies, T.B. (1883). On the growth of cork oak in Auckland. Transactions of the New Zealand Institute 15, 267–268.

Rheum palmatum L.

Polygonaceae

Common
Medicinal rhubarb, Turkey rhubarb, Chinese rhubarb.

Notes
A large herbaceous perennial native to mountain forest areas of Southwestern China and Tibet. Plants are reportedly grown for 6 to 10 years before being dug and the roots peeled and dried to produce medicinal rhubarb. Cultivation is centred in the Szechwan province of China at altitudes of 3 to 4 000 metres (Wallis, 1967). It is marketed in three grades, 'Shensi', 'Canton' and 'High–dried'. The first sort is assumed to be superior. A small trial planted at Lincoln in 1980/81 died out after twelve months but a further trial has been laid down by MAF at Tara Hills (M. Douglas) in 1987. *R. palmatum* is the only species now listed as official in the British Pharmacopoeia (1973). But at one time, *R. officinale,* Baill. was an alternative and this species is still cultivated on a small scale in Suffolk and is marketed as 'English rhubarb'. Other species listed in the Japanese Pharmacopoeia (1982) are *R. tanguticum* Maxim, and *R. coreanum* Nakai. Hybrids between the last species and *R. palmatum* are cultivated in Japan. *R. emodi* Wall. ex Meissn. was cultivated as a substitute in India during World War 2 and is still official in that country (Vide: Wealth of India – Raw Materials Vol 9, New Delhi 1972. pp. 3–4).

References
ITC. (1982). Markets for selected medicinal plants and their derivatives. UNCTAD/GATT, Geneva. pp 156–158.

Dress, W.J. (1964). *Rheum palmatum.* Baileya 12(1), 28–30.

Wallis, T.E. (1967). Textbook of Pharmacognosy. Fifth edn., J. & A. Churchill, London. pp 365–372.

See also: Wilson E.H. (1906). Chinese rhubarb. The Chemist and Druggist, 69, 371–372.

Rhus verniciflua Stokes

Anacardiaceae

Common
Japanese laquer tree, Varnish sumach, Urushi (Japan).

Synonyms
Rhus vernicifera DC., *Rhus kaempferi* Sweet.

Notes
Tall (to 18m), slender deciduous trees endemic or cultivated over a wide area from Southern Japan to the Himalaya and throughout Southeast Asia. The poisonous fruit are ornamental and trees are at least partly deciduous. The laquer is tapped from the bark in a similar manner to rubber latex.

References
The Wealth of India, Raw Materials, Vol 9, CSIR, New Delhi, 1972. pp 22–23.

Chun, Woon–Yeung. (1921). Chinese economic trees. Commercial Press, Shanghai. pp 219–221.

Weston, W. (1982). Lacquer work in Asia and beyond. Redwood Burn, Trowbridge. 345 pp.

Anon (1983) New technique exposes fake laquer. New Scientist 98(1354), p 149.

Ricinus communis L.

Euphorbiaceae

Common
Castor bean, Castor oil, Erandi (Hindi), Lama papalagi (Samoa), Beleni–vavalagi (Fiji), Tangan–tangan (Philippines).

Notes
A short lived perennial shrub from the Old–World tropics now distributed throughout the warmer regions of the world. Principal areas of cultivation include Brazil and India. Though castor oil plants appear to grow well in many warm areas, including northern New Zealand and most Pacific Islands, cultivation is not simple. The species is markedly variable in the oil content of seeds. Hand harvesting is difficult and seeds must be decorticated to produce high grade oil. Castor oil is now little used in medicine but has wide use as a lubricant in specific engineering fields. The seeds are poisonous due to the presence of a glycoprotein 'ricin', which is not extracted with the oil.

References
El Baradi, T.A. (1969). Castor bean. Tropical Abstracts 24(9), 567–572.

Raymond, W.D. (1961). Castor beans as food and fodder. Tropical Science 3(1), 19–23.

See also: Connor, H.E. (1977). The poisonous plants of New Zealand. 2nd edition, Government Printer, Wellington. pp 78–80.

Rorippa nasturtium–aquaticum (L.) Hayek

Cruciferae

Common
Water cress.

Synonyms
Nasturtium officinale R.Br., *N. aquaticum* .

Notes
A widespread aquatic plant on both main islands of N.Z. except for Taranaki Province. However, the commonest species in the South Island is the tetraploid *R. microphylla* which has little commercial value. Watercress was introduced by the French to Akaroa in 1840 and spread rapidly. Little use is made of the plant in New Zealand but it is cultivated in specially prepared beds in the United Kingdom and Germany.

References
Lyon, H.G. and Howard, H.W. (1952). Science and watercress growing. Journal of the Ministry of Agriculture 59, 123–128.

Ryder, J.E. (1979). Leafy salad vegetables. Avi Publishing Co., Westport. pp 240–248.

Palmer, J. (1988). A note on watercress. Commercial Herb Producers Association Newsletter 3(4), 4–5.

Rosa damascena Mill.

Rosaceae

Common
Damask rose, Rose of Kazanlik.

Notes
The principal rose cultivated in Bulgaria and India for the production of otto or attar of roses widely used in the perfume industry. The perfume is derived from the petals which tend to produce most at equable temperatures in the absence of wind. A plot of this species has been recently established at Riwaka Research Station. The species is thought to be of hybrid origin and the variety most grown in Bulgaria is 'Trigintipetala'. *R. x alba* L. is also grown in Bulgaria for scent and *R. x centifolia* near Grasse in France.

References
Windrlechner, M.P. (1981). The history and utilisation of *Rosa damascena.* Economic Botany 35(1), 42–58.

Singh, D.V. and Ram, M. (1987). Effect of spacing, extent of pruning, growth hormone and nutrients on flower yield of essential oil bearing rose (*R. damascena*) in sub–tropical India. Acta Horticulturae 208, 83–86.

N.B. A note on a possible new method of scent extraction from rose petals, not involving boiling, developed at the University of York. Vide: Anon (1987). Nose for a rose. Country Life. 10th September 1987.

Saccharum edule Hassk.

Gramineae

Common
Fiji asparagus, Duruka (Fiji), Lowland pitpit (New Guinea).

Synonyms
Erianthus maximus Brongn. var. *abortive* Grassl.

Notes
Possibly a sterile form of *Saccharum robustum,* a tall grass native to lowland Papua New Guinea and possibly some islands of Melanesia. The swollen aborted flower shoots are collected in season and are a popular vegetable. In Fiji there are several recognised cultivars. The species was introduced to Malaysia at an early period but is unimportant in that region.

References
Parham, B.E.V. (1948). Duruka, an edible flower. Agricultural Journal, Fiji 19(1), 16.

Powell, J.M. (1982). Traditional vegetables in Papua New Guinea. In: Bourke and Kesavan (eds)., Proceedings of the 2nds P.N.G. Food Crops Conference. Department of Primary Industry, Port Moresby. pp 64–77.

N.B. For discussion of specific status of this species see: Parham, J. (1979). Poaceae. In: A.C. Smith, Flora Vitiensis Nova. Vol. 1, pp 370–371.

Sagittaria trifolia L. *var. edulis* (Sieb.) Ohwi

Aponogetonaceae

Common
Arrowhead, Kuwai (Japan), Taro tinito (Tahiti), Tsz'ku (Chinese, Tahiti).

Synonyms
Sagittaria sinensis Sims, *S. sagittifolia* L. vars.

Notes
An aquatic herb bearing small starchy tubers at the apex of stolons. Native and occasionally cultivated in the Far East. Introduced to Tahiti and Moorea by Chinese market gardeners.

References
Porterfield, W.M. (1940). The arrowhead as a food among the Chinese. Journal of the New York Botanic Gardens 42, 35–47.

Herklots, G.A.C. (1972). Vegetables in Southeast Asia. George Allen and Unwin, London. pp 494–496.

Santalum acuminatum (R.Br.) A.DC.

Santalaceae

Common
Quandong.

Notes
A semi parasitic Australian shrub with edible fruit. Native to arid inland areas.

References
Sedgley, M. (1982). Preliminary assessment of an orchard of quandong seedling trees. Journal of the Australian Institute of Agricultural Science 481, 52–56.

Passingham, J. (1986). Selection for a better quandong. Australian Horticulture 84(2), 55–59.

Santalum album L.

Santalaceae

Common
Sandalwood, Sandal (Hindi), Yasi (Fiji), Laau ala (Hawaii), Ahi (Tahiti). NB; the Pacific Island names refer to allied species (see list below).

Notes
An evergreen tree from southern India to Indonesia. Parasitic in the early stages, requiring exposure to light for full development. The heartwood and roots are markedly aromatic and have a long history of use in the Far East for carving and incense. Now much of the wood is used for distilling sandalwood oil. A further 16 species of Santalum have been exploited in the Pacific region (see list below).

References
Sen–Sarma, P.K. (1982). Sandalwood – its cultivation and utilisation. in: Atal and Kapar (eds). Cultivation and utilisation of aromatic plants. CSIR, Jammu–Tawi, pp. 395–405.

The wealth of India – Raw materials. Vol. 9. CSIR, New Delhi 1972. pp 208–224.

N.B. For a revue of the 19th century exploitation of sandalwood in the Pacific see:

St John H. (1947). The history, present distribution and abundance of sandalwood on Ohau, Hawaiian Islands. Pacific Science 1(1), 5–20.

St John (1947) lists the following species of *Santalum* as yielding commercial sandalwood:

Papua New Guinea – *S. magregorii, F. Muell., S. papuanum* Summerh.

Australia – *S. lanceolatum* R.Br., *S. obtusifolium* R.Br. *S. ovatum* R.Br.

New Caledonia and New Hebrides – *S. austrocaledonicum.*

Fiji and Tonga – *S. yasi* Seem.

French Polynesia and Mitiaro (Cook Is.) *S. insulare* Bertero.

Henderson Island – *S. hendersonense* F.B.H. Br.

Hawaii – *S. ellipticum* Gaud., *S. freycinetianum* Gaud., *S. haleakalae* Hillebr., *S. lanaiense* Rock, *S. paniculatum* H & A., *S. pilgeri* Rock, *S. pyrularium* Gray.

Bonin Islands – *S. boninense* (Nakai) Tuyama.

Sapium sebiferum Roxb.

Euphorbiaceae

Common
Chinese tallow tree.

Notes
A medium to large (to 18m) deciduous tree native to south China, just frost hardy but more suited to the higher rainfall areas of the sub tropics. Vegetable tallow is obtained from the waxy endosperm covering the seeds. The seed kernels are also crushed to yield 'stilingia oil'. Vegetable tallow is used for cosmetics, candles and soap, and if properly distilled is edible.

References
Howes, F.N (1949) The Chinese tallow tree *Sapium sebiferum* Roxb.. Kew Bulletin (4), 573–580.

Duke, J.A. (1989). Handbook of nuts. CRC Press, Boca Raton. pp 262–264.

The Wealth of India, Raw materials, Vol 9, CSIR. New Delhi, 1972. pp 230–231.

Scheld, H.W. (1981). Woody biomass potential of the Chinese tallow tree. Economic Botany 35(4), 391-397.

Sauropus androgynus (L.) Merr.

Euphorbiaceae

Common
Star gooseberry(in India), Chekur manis (Malay).

Synonyms
Sauropus albicans Blume, *Phyllanthus strictus* Roxb.

Notes
Small evergreen shrubs. The leaves are used as spinach in India and Southeast Asia. The dull white acid fruit with dark red persistent calyx are made into chutney in Indonesia.

References
Ramachandran C. et al. (1980). Chekurmanis a multi–vit leafy vegetable. Indian Horticulture 25(1), 17.

Ochse, J.J. (1977). Vegetables of the Dutch East Indies. Australian National University Press, Canberra. pp 290–292.

Saussurea lappa C.B. Clarke

Compositae

Common
Costus, Kuth, Kut (Hindi).

Synonyms
Aucklandia costus Falc.

Notes
A tall herbaceous perennial from forest margins in the dry western Himalayas. The bitter fragrant root has long been part of traditional 'Eastern' medicine and is still included in the Japanese pharmacopoeia (1982). Costus has been an Indian export for centuries but has recently been declared an endangered species. N.B. Mu–hsiang and Mokko are respectively the Chinese and Japanese names for the crude drug which is the dried root of this species.

References
The Wealth of India – Raw Materials Vol. 9, CSIR, New Delhi, 1972. pp 240–244.

Arora, R.K. et al. (1982). Genetic diversity in the cold arid western Himalayas. Plant Genetic Resources Newsletter 51, 27–30.

Schinus terebinthifolius Raddi

Anacardiaceae

Common
Broadleaf pepper tree, Pink pepper, Brazilian pepper, Florida holly, Christmas berry (Hawaii), Warui (Fiji).

Notes
A medium sized tree, indigenous to the coastal forests of tropical Brazil. Spread throughout tropical areas, including northern New Zealand and Fiji as an ornamental tree. The pink berries are poisonous but have been used as a minor spice.

References
Morton, J.F. (1978). Brazillian pepper – its impact on people, animals and the environment. Economic Botany 32(4), 353–359.

Stahl, E. et al. (1983). Cardanol, a skin irritant in pink pepper. Planta Medica 48, 5–9.

Scorzonera hispanica L.

Compositae

Common
Scorzonera, Black salsify.

Synonyms
S. denticulata Lam., *S. edulis* Moench.

Notes
A perennial herb with edible roots. Usually cultivated as a biennial. From southern Europe but grown commercially further north. Rare in Pacific region.

References
Scheffer, J. (1984). Taking the shell off the vegetable oyster. Southern Horticulture No 17: 51–52

Scheffer, J. (1983). The culture of scorzonera or black salsify. Information Sheet. MAF, Cronins Road, Pukekohe. 5 pp.

Sechium edule Swartz

Cucurbitaceae

Common
Chayote, Choko, Milton.

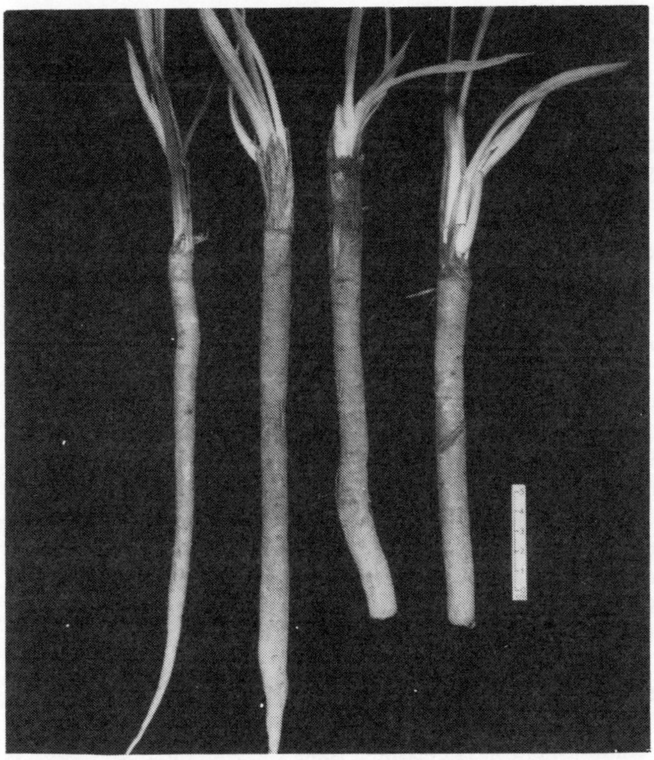

Figure 51: Scorzonera.

Notes

An easily cultivated gourd with a single large seed. Originally from South America, but now widely cultivated in warm temperate and tropical regions for both its fruit and the shoot tips. Unimportant but widespread in northern New Zealand and many Pacific Islands.

References

Adams, C.D. (1968). The chayote, milton or vegetable pear. American Horticulture Society Magazine 47(3), 314–315.

Morton, Julia (1967). The balsam pear – an edible, medicinal and toxic plant. Economic Botany 21(1), 57–68.

Vide also: Anon (1983). Chokos need three things. Southern Horticulture No. 10, p 13.

Sesamum indicum L.

Pedaliaceae

Common

Sesame, Gingelly, Beni (seeds), Goma (Japan), Lenga (Malay), Linga (Philippines).

Synonyms

Sesamum luteum Retz., *S. occidentale* Heer. et Reg., *S. oleiferum* Moench., *S. orientale* L.

Notes

An erect annual herb with an indeterminate flower spike (c.f. foxgloves). The seed capsules of all traditional cultivars shatter making harvesting difficult. There are two main groups of sesame with light or dark coloured seeds. Plants are adapted to warm, dry conditions and test plots of Indian cultivars grown at Otara Research Station in the mid 1970's showed cold symptoms. However, sesame is grown in Korea and cultivars from this region could be more cold tolerant.

References

Tribe, A.J. (1967). Sesame. Field Crop Abstracts 20(3), 189–194.

Ashri, A. (1985). Sesame and safflower: status and potentials. FAO, Plant Production and Protection Paper No. 66, Rome.

Claxton, R.A. (1972). Review of sesame. Australian Specialist Conference on crops of potential economic importance. Sydney, 1972, 3:46 – 50.

See also: FAO Sesame and safflower newsletter (A. Ashri ed.) No. 1 Rome, 1985.

Simmondsia chinensis (Link) Schneider

Simmondsiaceae

Common

Jojoba, Goat nut.

Synonyms

Simmondsia californica Nutt.

Figure 52: Sesame, Otara Research Station 1975.

Notes
A dioecious desert shrub from the Sonoran desert area of Mexico and southwestern United States. The seeds yield a liquid wax which is a substitute for sperm whale oil. Plants are winter flowering, so experimental planting in New Zealand and southern regions of Australia have been adversely affected by frost. An experimental planting at Lincoln did not survive the first winter.

References
National Academy of Sciences (1975). Products from jojoba: a promising new crop for arid lands. NAS, Washington. 30 p.

Eaton, P. (1985). Jojoba in perspective 12 years on. Australian Horticulture, March 1985. pp 70–77.

Sium sisarum L.

Umbelliferae

Common
Skirret, Chervin, Crummock (Scotland), Mukago–ninjin (Japan).

Synonyms
Carum sisarum Baill., *Sium ninsi* L.

Notes
The species is native to an extended area from eastern Europe and southeastern Russia to east Asia and Japan. However, the roots are markedly more developed in the cultivated skirret than in feral European plants and in Japan plants are characterised by the presence of bulbils in leaf axils which do not appear in cultivars present in New Zealand. The roots are sweet (some 6% sucrose), but retain a central fibrous core and no 'improved' cultivars are known. This root crop was probably introduced to Europe via Germany as early as the 12th century but does not appear to be the 'siler' sent to Rome as annual tribute to Tiberius (Pliny, Historia Naturalis, liber XIX cap. 5). In the 17th century skirret was a popular vegetable in the United Kingdom but though Miller (Gardeners Dictionary, 1731) described it as "wholesemeth and nourishing roots" he goes on to say "yet it is at present very rare to meet with it in the gardens near London". A rarity it has remained ever since.

References
Anon. (1899). Skirret (*Sium sisarum* L.). Kew Bulletin 443, 39–42.

Kennelly, A.G. (1952). The home vegetable garden. N.Z. Department of Agriculture, Bulletin 243, Wellington. p 152.

Solanum aviculare Forst f.

Solanaceae

Common
Poroporo (New Zealand), Kangaroo apple (Australia).

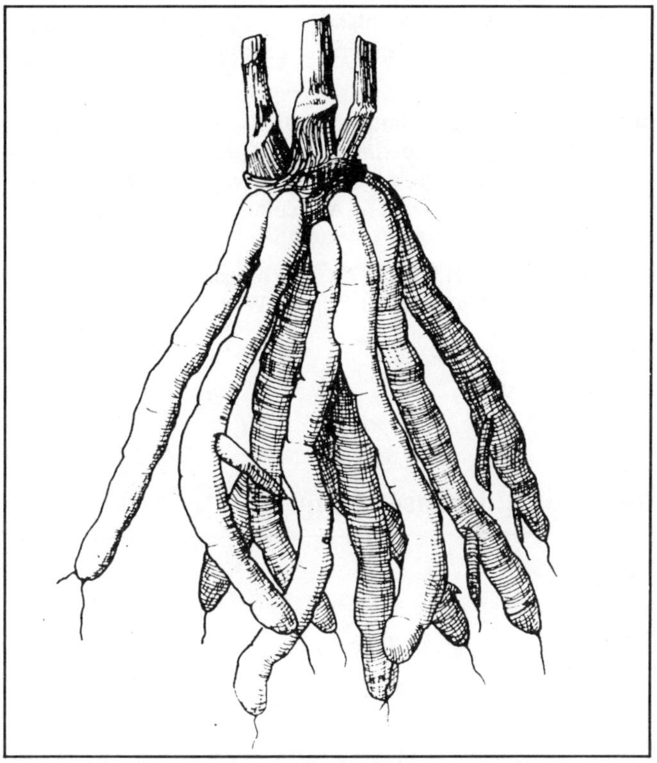

Figure 53: Skirret root.

Notes
A shrub naturally occuring in bush margins from northern New Zealand through eastern Australia to the New Guinea highlands. In the South Island of New Zealand this species is replaced by the hardier and larger tetraploid *S. laciniatum* Ait. Both species were cultivated during the decade 1965–75 as a source of the alkaloid 'solasodine', a steroid drug precursor. However, alkaloid yields from the tetraploid species proved uneconomic and the whole programme was discontinued following the collapse of the pharmaceutical industry based on Mexican yams. For a commentary on this see Djerassi (1976).
S. laciniatum is still being tested in India (Bharati and Chatterjee,1987).

References
Mann, J.D. (1978). Production of solasodine for the phamaceutical industry. Advances in Agronomy 30, 207–245.

Alspach, P.A. (1986). Poroporo (*Solanum aviculare* Forst. f.): a summary and discussion of some recent agronomic research in New Zealand. New Zealand Journal of Experimental Agriculture 11(1), 11–16.

See also:

Anon (1971). From weed to wealth. N.Z. Journal of Agriculture 123(1), 15–16.

France, J.T. (1971). Growth of the steroid drug industry. N.Z. Journal of Agriculture. 123(1), 17.

Djerassi, C. (1976). The manufacture of steroidal contraceptives: technical versus political aspects. Proceedings of the Royal Society (London) B. 195, 175–186.

Bharati, P. an Chatterjee, S.K. (1987). Agroeconomics of *Solanum laciniatum* Ait. – a new medical plant in Darjeeling hills. Acta Horticulturae 208, 253–260.

Solanum khasianum Clarke

Solanaceae

Common
Khasianum.

Synonyms
Solanum viarum Dunal.

Notes
A spiny perennial from northeastern India grown as a source of solasodine (cf. our native *S. aviculare* Forst f.). Trials at Otara Research Station 1973–76 indicated that relatively high yields of alkaloids were obtainable from the berries and that this had not depreciated during storage. Currently this species is being grown commercially in India.
N.B. Solasodine is a nitrogen analogue of diosgenin, usually obtained from yams, and a precursor for steroid drugs.

References
Yaniv, Z. et al. (1984). The effect of moisture stress on growth and solasodine content in *Solanum khasianum.* Planta Medica 50(1), 60–65.

Palmer, J. (1977). Trials with *Solanum* species. New Zealand Journal of Horticulture 1, 18–21.

Solanum melongena L.

Solanaceae

Common
Aubergine, Eggplant, Brinjal, Baigun (Hindi), Torong (Malay), Talong (Philippines).

Synonyms
Solanum esculentum Dun., *S. longum* Roxb.

Notes
A short lived shrubby perennial usually cultivated as an annual in warm temperate regions. In India feral plants are often spiny and the fruit bitter. Cultivars are invariably bland tasting and appear to be self pollinated. A very popular vegetable in India and the cultivation in Fiji is largely in the hands of Indian migrants. However, in Tahiti aubergines are grown by Chinese market gardeners. Aubergines have been known in China since the 5th century B.C. and were probably introduced to Europe by the Arabs in the 15th century A.D. However, they were not widely grown before the 17th. In New Zealand, aubergines are essentially a glasshouse crop and remain relatively unknown.

References
Martin, F.W. and Pollack B.L. (1979). Vegetables for the hot humid tropics. Part 5, Eggplant, *Solanum melongena.* U.S. Department of Agriculture, New Orleans. 18 pp.

Warde, J. (1985). Madde apples. Being an account of eggplant. Organic Gardening 32 (4), 30–38.

Solanum nigrum L.

Solanaceae

Common
Black nightshade, Garden huckleberry, Oupoo (Tahiti), Magalo (Samoa), Boro ni veiwere (Fiji), Makoiya (Fiji Indian), Poro (Rarotonga), Polopolo (Hawaii).

Synonyms
S. vulgatum Willd.

Notes
A variable cosmopolitan species of annual herb. Several botanical varieties have been described including 'var. guineense' which is one of the huckleberries of the U.S.A. and ocasionally cultivated in New Zealand gardens. In New Zealand feral black nightshade is considered a weed, but there has been confusion in the past with *S. americanum* Miller, the small–flowered black nightshade, a very similar species widespread in the Pacific Islands, throughout Southeast Asia and in the warmer parts of New Zealand. The young shoots of this species are gathered as a spinach substitute in many Pacific Islands and Indonesia. The berries are also eaten.

References
Degener, O. (1945). Plants of Hawaii National Park illustrative of plants and customs of the South Seas. Edwards Brothers, Ann Arbor. pp 257–261.

Heiser, C.B. (1969). Nightshades. The paradoxical plants. W.H. Freeman, San Francisco. pp 63–105 (for huckleberry).

Fortuin, F.T.J.M. and Omta, S.W.P. (1980). Growth analysis and shade experiment with *Solanum nigrum* L., the black nightshade, a leaf and fruit vegetable in West Java. Netherlands Journal of Agricultural Science 28(4), 199–210.

See also: Connor, H.E. (1977). The poisonous plants of New Zealand. Government Printer, Wellington. pp 169–170.

Solanum quitoense Lam.

Solanaceae

Common
Naranjilla.

Notes
A shrubby perennial with large soft hairy leaves with a purple sheen. Very frequently spiny but there are unarmed cultivars. The fruit is yellow – orange skinned and green fleshed, a little larger than a hens egg. In South America naranjilla is used as fruit juice. In New Zealand it seems destined to remain as an occasional ornamental in the frost free north.

References
Heiser, C.B. (1985). Ethnobotany of the naranjilla and its relatives. Economic Botany 39(1), 4–11.

Dennis, F.G. et al. (1985). Naranjilla: a potential cash crop for the small farmers of Latin America. Acta Horticulturae 158, 475–481.

Solanum repandum Forst. f.

Solanaceae

Common
Island tomato, Taulo'u (Samoa), Boro–sou (Fiji), Loka moka (Niue).

Notes
A small woody perennial (to 2 m) with large pubescent leaves. The fruit was at one time cooked and eaten in some Polynesian islands and in Fiji (vide Parham, 1942). Plants are now rare but widespread in the South Pacific. This is one of the species tried at Koronivia Research Station (Fiji) as a possible rootstock for the common tomato during the mid 1970's. Some plants were also grown at Otara over the same period but failed to ripen any fruit.

References
This species has been largely ignored since the 19th century and it was not included in the US wartime emergency bulletin 'Emergency food plants and poisonous plants of the islands of the Pacific' by E.D. Merrill (Washington, 1943) or the New Zealand equivalent, 'Food is where you find it' by Lucy Cranwell et al. (Auckland Institute and Museum, 1943).

An excellent drawing (plate 38) and some notes are included in: Seeman, B. (1865). Flora Vitiensis. Lovell Reeve and Co., London. p 176.

See also: Brown, F.B.H. (1935). Flora of Southeastern Polynesia. III Dicotyledons. B.P. Bishop Museum, Bulletin 130, Honolulu. p 257.

Parham, B.E.V. (1942). Some useful plants of the Fiji Islands. Agricultural Journal, Fiji 13 (2), 39–47.

Solanum viride R. Br.

Solanaceae

Common
Cannibals' tomato, Borodina (Fiji), Porohiti (Rarotonga), Uporo (Tahiti).

Synonyms
Solanum anthropophagorum Seem., *Solanum uporo* Dunal.

Notes
A shrubby perennial often about 1.5 m high with large dark green ovate leaves and bright red fruit. The common English name derives from the past practice in Fiji of wrapping human flesh in the leaves of this species prior to cooking. According to H.B.R. Parham (1943) the leaves of this species make a good pot herb and the fruit is recorded by Seemann (1862) as occasionally being used in a sauce. The faintly aromatic fruit have also been used to adorn ceremonial leis in the Marquesas (Brown, 1935 opp. cit.). Borodina were used at Koronivia Research Station (Fiji) in the mid 1970's as a rootstock for the common tomato. Plants grown at Otara Research Station over the same period did not ripen any fruit.

References
The original reference to *Solanum uporo* usage in Fiji was;

Seemann, B. (1862). Viti: an account of a government mission to the Vitian or Fijian Islands. Macmillan & Co., Cambridge. p 177.

Anon (1864). The cannibals' tomato. Gardeners' Chronicle, 9 January 1864. pp 27–28. This was based on Seemann's 1862 report to the British Parliament.

Most subsequent notes on the utilisation of *S. uporo* seem to have been derived from these sources, e.g.

Degener, O. (1949). Naturalists' South Pacific Expedition: Fiji. Paradise of the Pacfic Ltd, Honolulu. Chapter 16, 'Cannibalism'. pp 155–176.

Seemann's plate of *S. uporo* was reproduced in: Parham, J.W. (1964). Plants of the Fiji Islands. Government Press, Suva. p 238.

Sorghum bicolor (L.) Moench

Gramineae

Common
Sorghum, Guinea corn, Jola or Jawa (India), Kaoliang (China), Doura (Tahiti), Tohuka (Maori).

Synonyms
Holcus sorghum L.; *Andropogon sorghum* (L.) Brot., *Sorghum vulgare* Pers.

Notes
Sorghum is the 4th ranking world cereal and would not be included in these notes except for the fact that it is relatively unknown in the Pacific area. It is a crop adapted to regions with high temperatures and uncertain rainfall. There are three principal groups of cultivars:
Grain sorghums – the grain is frequently ground but the flour quickly becomes rancid if the embryos are not removed. 'Pearled' grain may be cooked like rice.

Broomcorn (var. *dochna* Forsk.) – sorghum inflorescences where the rachis is much shortened but long side branches make useful bristles. Main areas of production include the U.S.A. and Fiji. Trials were conducted in Auckland (PDD/DSIR) in the early 1970's.

Sugar sorghums – the stems are a source of sucrose.

References
Purseglove, J.W. Tropical crops – Monocotyledons. Longman, London 1972. pp 259–287.

Figure 54: Sorghum.

ICRISAT (1982). Sorghum in the eighties. Proceedings of the International Symposium on sorghum, Patancheru 1981. 2 volumes.

See also:

Hemmerly, T.E. (1983). Traditional method of making sorghum molasses. Economic Botany 37(4), 406–409.

Gillies, T.B. (1882). On sorghum experiment. Trans. N.Z. Institute 14, 373–374. Idem (1883). 15, 261–267. Idem (1884). 16, 512–517.

Anon (1986). Sorghum as a sugar source. Growing Today 3(4), 36.

Thompson P.G. (1966). Growing broom corn. The Fiji Farmer 2(3), 46–48.

Spondias cytherea Sonn.

Anacardiaceae

Common

Otaheite apple, Ambarella, Polynesian plum, Wi (Fiji), Vi (Polynesia), Aura (Fiji Indian).

Synonyms

Spondias dulcis Parkinson.

Notes

Large deciduous tree probably indigenous to the South Pacific. The fruit is locally popular but contains a large fibrous stone. N.B. The related hog plum *Spondias pinnata* Kurz is cultivated in Indonesia, the Philippines and New Guinea. The fruit is inferior and smaller.

References

Uhe, G. (1974). Wayside plants of the South Pacific. Stocton House, Auckland. pp 123–124.

Allen, Betty M. (1967) Malayan fruits. Donald Moore Press, Singapore. pp 20–23.

Sturrock, D. (1959). Fruits for southern Florida. Southeastern Printing Co., Stuart. p 52.

Stachys affinis Bunge

Labiatae

Common

Chinese artichoke, Knotroot, Crosnes (France), Choro–gi (Japan), Kon–lo (China).

Synonyms

Stachys sieboldii Miq., *S. tubifera* Naud.

Notes

A small tuber crop introduced to Europe from China in the mid 19th century but only really became established in France. A virus free clone was brought to N.Z. from Vilmorin Andrieux, Paris in 1980 (vide: J Palmer, Lincoln). Tubers require little cooking and are often pickled in Japan.

References

Herklots, G.A.C. (1972). Vegetables in South east Asia. Allen and Unwin, London. pp 470–471.

Porterfield, W.M. (1951). The principal Chinese vegetable foods and food plants of Chinatown markets. Economic Botany 5(1), 20–23.

Figure 55: Chinese artichokes.

Peron, J–Y. (1980). Le crosne du Japon (*Stachys sieboldii* Miq.) Revue Horticole 210, 11–17.

Stevia rebaudiana Bertoni

Compositae

Common
Stevia, Azucaca (Paraguay).

Synonyms
Eupatorium rebaudianum Bertoni

Notes
An herbaceous perennial from the Andean region of Paraguay. Cultivated for the leaves which contain the alkaloid 'stevoside' used as a sweetening agent. At least two introductions made to N.Z. in the early 1980's. Trials at Levin (vide Les Tilbury) indicate little prospect of commercial exploitation.

References
Nichols, Rae (1985). The case of the sugar cane. Dittany 6, 56–58.

Soejarto, D.D. et al. (1983) Potential sweetening agents of plant origin, 2. Field search for sweet tasting *Stevia* species. Economic Botany 37(1), 71–79.

See also: Planas, G.M. and Kuc, S. (1968). Contraceptive properties of *Stevia rebaudiana.* Science 162, 1007.

Stokesia laevis (Hill) Greene

Compositae

Common
Stokesia, Stoke's aster.

Synonyms
Carthamus laevis Hill.

Notes
An erect, branched perennial (to 70 cm) from southeastern United States. Long grown in Europe for its ornamental violet blue flowers and there are named cultivars in the 'trade'. Stokesia occur in New Zealand gardens but are comparitively uncommon. In the 1970's there was some interest in this species as a potential oil seed crop. Seeds contain some 40% oil of which 70% is epoxyoleic acid used as a stabiliser in vinyl plastics.

References
Gunn, C.R. and White, G.A. (1974). *Stokesia laevis:* taxonomy and economic value. Economic Botany 28(3), 130–135.

Symphytum officinale L.

Boraginaceae

Common
Comfrey, Boneset, Knitbone, Consolida.

Notes
An erect perennial herb, native to damp areas of Europe. The flower colour varies from white to purple. Both leaves and roots are used in herbal medicine particularly for gastric ulcer. But, as its common names suggest, it has also been used as a wound dressing. The active ingredient allantoin, has been known to cause liver damage in overdose.
N.B. Other species of *Symphytum* in New Zealand occurring as garden escapes:
S. asperum Lepech. Rough comfrey or prickly comfrey. Originally from southern Russia to Iran. Introduced as a forage plant. The flowers change from rose to clear blue.
S. x uplandicum Nym. (syn. *S. peregrinum* Ledeb.) Russian confrey. Thought to be a hybrid, *S. officinale x S. asperum.* The flower colour changes from rose to purple with maturity.

References
Hills, L.D. (1976). Comfrey. Past, present and future. Faber & Faber, London. 253 pp.

British Herbal Pharmacopoeia (1983). British Herbal Medicine Association, Cowling. pp 201–203.

Crawford, J.C. (1877). Some remarks on the plant called prickly comfrey. Transactions of the N.Z. Institute 10, 523–524,528.

Syzygium malaccense (L.) Merr. and Perry

Myrtaceae

Common
Malay apple, Kavika (Fiji), Kahika (Cook Is.), Lahika (Solomon Is.), Fekika (Tonga), Fekakai (Niue), Nonu fi'a fia'a (Samoa), Ahia (Tahiti).

Synonyms
Eugenia malaccense L., *Jambosa malaccensis* (L.) DC.

Notes
A small tropical tree (to 13m) with sweet watery, pear shaped fruit. Native to Southeast Asia but widely naturalised and cultivated in the islands of the South Pacific. Probably an aboriginal introduction to this region. Some other *Syzygium* species are native there, including; *S. maire* (Cunn.) Sykes et Garnock–Jones, in New Zealand.
Other introduced species cultivated on Pacific Islands include; *S. cuminii* (L.) Skeels. (Syn. *Eugenia jambolana* Lam., *S. jambolanum* (Lam.) DC.). The Java plum, Jambolan (Malay), Duhat (Philippines), Jamun (Hindi), Pistas (Tahiti), Kahika papaa (Cook Is., Kavika ni India (Fiji).
S. jambos (L.) Alston, (syn. *Eugenia jambos* L., *Jambosa vulgaris* DC.). Rose apple, Kavika ni vavalangi (Fiji), Kahika papaa (Cook Is.), Ahia popaa (Tahiti), Seasea papalanga (Samoa).
S. australe (Link) B. Hyland, the Australian brush cherry is cultivated in some gardens of northern New Zealand (Gardner, 1987).
Dried flower buds of *S. aromaticum* (L.) Merr. form the spice 'cloves'.

References
Morton, J.C. (1987). Fruits of warm climates. Creative Resource Systems Inc., Winterville. pp378–381.

Allen B.M. (1967). Malayan fruits. Donald Moore Press, Singapore. pp121–125.

Petard, P. (1986). Plantes utiles de Polynesie. Editions Haere Po No Tahiti. pp242–243.

See also: Gardner, R.O. (1987). *Syzygium* and related genera (Myrtaceae) in Auckland. Auckland Botanical Society Newsletter 42(1), 12–14.

Tacca leontopetaloides (L.) Kuntze

Taccaceae

Common
East Indian arrowroot*, Polynesian arrowroot, Pia (Polynesia), Yabia (Fiji).

Synonyms
Tacca pinnatifida Forst. f., *T. involucrata* Schum. et Thonn.

Notes
Indigenous to the eastern tropics. At one time cultivated in Tahiti and Hawaii. Now seldom grown but naturalised in many Pacific Islands. Grows best in partial shade.

References

Massal, E. and Barrau, J. (1956). Food plants of the South Sea islands. South Pacific Commission, Technical Paper No 94, pp 11–12.

Kay, D.E. (1973). TPI Crop and Product Digest, No.2. Root Crops. Tropical Products Institute, London. pp 57–60.

*vide *Maranta arundinacea* – West Indian arrowroot.

Talinum triangulare Willd.

Portulacaceae

Common
Surinam purslane, Potherb fame flower, Sweet heart, Water leaf, Talinum.

Synonyms
Portulaca paniculata L., *Talinum crassifolium* Willd.

Notes
An erect branching herb characterised by three angled peduncles. The broad, dark green fleshy leaves may be used fresh in salads or cooked as a spinach substitute. Talinum is native to central America but is introduced and cultivated in Indonesia and the Philippines. Another species *T. paniculatum* (Jacq.) Gaertn. often grown in the U.S.A. as an ornamental and occasionally in New Zealand. It has also been introduced to Fiji (called Varataisi), Hawaii,and the Sydney region of New South Wales. The two species are sometimes confused.

References

Ochse, J.J. (1977). Vegetables of the Dutch East Indies. Australian University Press, Canberra. pp 620–621.

Herklots, G.A.C. (1972). Vegetables in Southeast Asia. George Allen and Unwin, London. pp 161–162.

Tanacetum cinerariifolium (Trev.) Schultz-Bip.

Compositae

Common
Pyrethrum.

Synonyms
Chrysanthemum cinerariifolium (Trev.) Vis.

Notes
A mountain daisy from Southeastern Europe cultivated at high altitudes in East Africa for the flower heads which contain a natural insecticide. Trials were conducted in New Zealand in the 1930's and 1940's and again in the 1970's but the pyrethroid content of flowers was low. Dr Bhat in Tasmania has reportedly produced cultivars with a higher yield potential (Crop Science 1984, 24, p 619).

References

UNCTAD/GATT. Pyrethrum a natural insecticide with growth potential. Geneva, 1976. 113 pp.

Anon (1986). Pyrethrum in Tasmania. Australian Horticulture 84(4), 10.

Keen M.(1988) Natural pyrethrum makes a comeback. Ceres 21(3), 6–7.

N.B. For original N.Z. trials see:
Chamberlain, E.E. and Proctor, C. (1947). N.Z. J. Sci. Technol. 28A(6), 353–361.

Chamberlain, E.E. and Clark, P.J. (1948). J. Sci. Technol. 29A (4), 215–222.

Clark, P.J. et al. (1948). J. Sci. Technol. 29A (4), 223–232.

See also: The Press, Christchurch 13/1/84. Lecturer leaves legacy of pyrethrum industry.

Tanacetum parthenium (L.) Schultz- Bip.

Compositae

Common
Feverfew.

Synonyms
Chrysanthemum parthenium (L.) Bernh., *Leucanthemum parthenium* (L.) Gren. and Godron.

Notes
A short (usually under 60 cm), branched perennial herb with yellowish green strongly aromatic leaves. The flowers are small (c. 2 cm diam.) but numerous with white ray florets surrounding yellow disc florets in the centre. Feverfew is common on both islands of New Zealand, in gardens and as an occasional escape. It has recently been demonstrated as a possible cure for migraine attacks.

References
Berry, M.I. (1984). Feverfew faces the future. The Pharmaceutical Journal, 18 May 1984. pp 611–614.

Johnson, E.S. et al. (1985). Efficacy of feverfew as a prophylactic treatment of migraine. British Medical Journal 291, 569–573.

Taraxacum kok–saghyz Rodin

Compositae

Common
Russian dandelion, Kok–saghyz.

Synonyms
Taraxacum bicorne Dahlst.

Notes
A plant closely related to the common dandelion from Russian Turkestan. Cultivated in N.Z. during World War 2 as a source of rubber, but not recorded since then.

References
Grimmet, R.E.R. (1942). Can rubber be grown in N.Z. New Zealand Journal of Agriculture 65(6), 325–326.

Zotov, V.D. and White, E.P. (1943). Rubber production in N.Z. Trials with kok–saghyz. New Zealand Journal of Agriculture 67(2), 75–78.

Telfairia pedata (Smith ex Sims) Hook.

Cucurbitaceae

Common
Oyster nut, Mkwene (Swahili).

Notes
A dioecious gourd from East Africa. Introduced into India and Southern China. The fruit is large (over 10 kg) with large flat seeds 2 to 3 cm wide which are eaten as nuts. The seeds are hard to crack.

References
Menninger, E.A. (1977). Edible nuts of the world. Horticultural Books, Stuart. pp 108–109.

Herklots, G.A.C. (1972). Vegetables of Southeast Asia. Allen & Unvwin, London. pp 344–346.

N.B. The related *Telfairia occidentalis* Hook.f.,is the fluted pumpkin of West Africa. This species has noted drought resistance and an ability to resprout following rain.

vide: Okodi, B.E. and Mgbeogu, C.M. (1983) Fluted pumpkin, *Telfairia occidentalis;* West African vegetable crop. Economic Botany 37(2), 145–149.

Terminalia catappa L.

Combretaceae

Common
Indian almond, Malabar almond, Talie (Samoa), Kauarika (Cook Is.), Autara'a (Tahiti), Tavola (Fiji), Ketapang (Malay).

Notes
A large tree of sandy shore sites in old world tropics. Widespread in the South Pacific and valued for the edible nuts. Difficulties in kernel extraction restrict commercial exploitation. N.B. *Terminalia kaernbachii* Warb. (syn *T. okari* C.T. White), is the 'Okari nut' of Papua New Guinea and adjacent islands. Okari nuts are described as cigar shaped, up to 6cm long, with concentric layers of slightly juicy white flesh. Accessions from the Galip area are said to be larger than others. Several other species of *Terminalia* are native to the islands of the South Pacific. vide: Fosberg, F.R. and Sachet, M.H. (1981). Polynesian plant studies 6–18. Smithsonian Contributions to Botany 47, 13–17.

References
Morton, Julia (1985). Indian almond *(Terminalia catappa* L). Salt tolerant, useful tropical tree with a nut worthy of improvement. Economic Botany 39(2), 101–112.

Uhe, G. (1974). Wayside plants of the South Pacific. Stocton House, Auckland. pp 127–128.

Menninger, E.A. (1977). Edible nuts of the world. Horticultural Books Inc., Stuart. pp 60–61,

Tragopogon porrifolius L.

Compositae

Common
Salsify, Vegetable oyster.

Notes
A European herb with an edible root. Cultivated in Europe since about 1600 but never of major importance. There are few known cultivars.

References

Scheffer, J. (1984). Taking the shell off the vegetable oyster. Southern Horticulture No.17, 51–52

Jankowiak, J. (1977). All the oysters don't grow in the ocean. Organic Gardening and Farming, April 1977, pp 156–159.

Trapa bicornis Osbeck.

Trapaceae

Common
Chinese water chestnut*, Ling Ko.
* Strictly speaking the name Chinese water chestnut refers to *Eleocharis dulcis* Trin.

Synonyms
Trapa bicornuta L.

Notes
An annual floating aquatic plant producing large edible seeds with two horns. This species was sold in New Zealand at least as early as 1923 by the Te Aro Seed Company of Wellington under the name 'Bull horn nuts'. It has not been recorded in this country since World War 2. In China the seeds are sold in autumn and must be cooked before eating as they are slightly toxic in the raw state. Once dried they quickly lose their viability. The nomenclature of this genus is confused. There are several species recorded from East and Southeast Asia which may or may not be sub–species of the above. Amongst the most important from the edible standpoint are:
T. bispinosa Roxb. The singhara nut of India and Sri Lanka.
T. japonica Flerov. The hishi of Japan, which occurs as far south as Taiwan.
T. natans L. The water caltrop, which has been consumed in Central Europe since prehistoric times. But its range extends into East Asia.

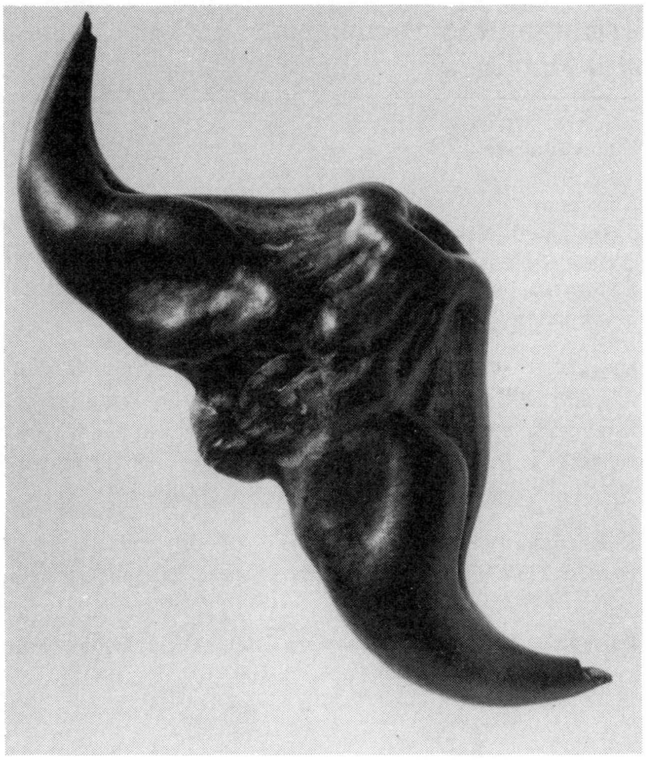

Figure 56: Trapa bicornis seed.

References

Porterfield, W.M. (1938). *Trapa bicornis,* a water chestnut known to the Chinese as Ling ko. Journal of the New York Botanic Garden 39, 134–138.

Herklots, G.A.C. (1972). Vegetables in Southeast Asia. George Allen and Unwin, London. pp 496–498.

See also: Mazundar, B.C. (1985). Water chestnut – the aquatic fruit. Cultivation in India. World Crops 37(2), 42–44.

Masefield, G.B. et al. (1969). The Oxford book of food plants. Oxford University Press, London. pp 32–33.

Trichosanthes cucumerina L. *var. anguina* (L.) Haines

Cucurbitaceae

Common
Snake gourd, New Guinea bean, Cucuzzi, Padwal (India).

Synonyms
Trichosanthes anguina L.

Notes
A gourd with long slender fruits which may exceed a metre in length. The species range is from New Guinea to North Africa. Nowhere is it important as a vegetable and it is only grown in the Pacific Islands or northern N.Z. as a peculiarity.

References
Dark, S. (1983) Have you tried the cucuzzi. Organic Gardening 30(4), 102.

Purseglove, J.W. (1968). Tropical Crops. Dicotyledons. Vol.1. Longman, London. pp 136–138.

Trigonella foenum–graecum L.

Leguminosae

Common
Fenugreek, Metha (Hindi).

Notes
An annual legume crop adapted to warm arid conditions. Used as a forage crop in southern Russia but the seed is an important constituent of curry powders in India and has been shown to be a source of diosgenin. Young shoots are eaten as greens and are sometimes seen in Auckland vegetable markets. Apart from India, fenugreek seed is exported from the Middle East and North Africa. Trial plantings at Lincoln in the early 1980's were just hardy but yields were low.

References
Seddon, Q. (1980). Fenugreek – spicy crop for forage and drugs. Farmer's Weekly 93(8), 16–17 and 21.

Duke, J.A. (1981). Handbook of legumes of world economic importance. Plenum Press, New York. pp 268–271.

Figure 57: Fenugreek. Note the long sickle shaped pods.

Tropaeolum tuberosum Ruiz. et Pav.

Tropaeolaceae

Common
Isanu (Bolivia), Mashua (Peru), Anu, Aniu (N.Z., probably a mispelling of anu.)

Notes
A small tuber crop of the Andean region. Introduced into New Zealand in the late 1970s (vide S. Dawes, Mt Albert). Mainly grown as an ornamental. A 'short–day' plant not flowering or producing tubers until autumn under N.Z. conditions. Flowers do not appear to set seed south of Blenheim.

References
Hodge, W.H. (1951). Three native tuber crops of the high Andes. Economic Botany 5(2), 185–201.

Townsend, J. (1964). Unexploited crops in Bolivia. World Crops 16(3), 67–68.

See also: Williams, L.O. (1978). The anu in Mexico. Economic Botany 32(1), 104.

Tylosema esculentum (Burchell) A. Schreiber

Leguminosae

Common
Marama bean, Marama (Tswana), Tsi or Tsin (Kung Bushmen).

Synonyms
Bauhinia esculenta Burchell

Notes
A scandent legume from arid areas of southern Africa producing a large edible root. Seed was brought to New Zealand in 1980 (D.J. Davies, CRD Lincoln) but plants did not survive the winter in the field.

References
Veitmeyer, N.D. (1978). The humble crops. Ceres 11(2), 23–27.

N.A.S. (1979). Tropical legumes: Resources for the future. National Academy of Sciences, Washington. pp 68–74.

Ullucus tuberosus Loz.

Basellaceae

Common
Ulluco, Ollocu.

Notes
An Andean tuber crop. One accession introduced into New Zealand c. 1978 (vide S. Dawes, DSIR, Auckland). Tubers are produced very late in autumn making the line difficult to maintain at Lincoln.

References
Hodge, W.H. (1951). Three native tuber plants of the high Andes. Economic Botany 5(2), 185–201.

Vietmeyer, N. (1984). The lost crops of the Incas. Ceres 17(3), 37–39.

Figure 58: Ulluco. Note tuber forming at the tip of axillary shoot.

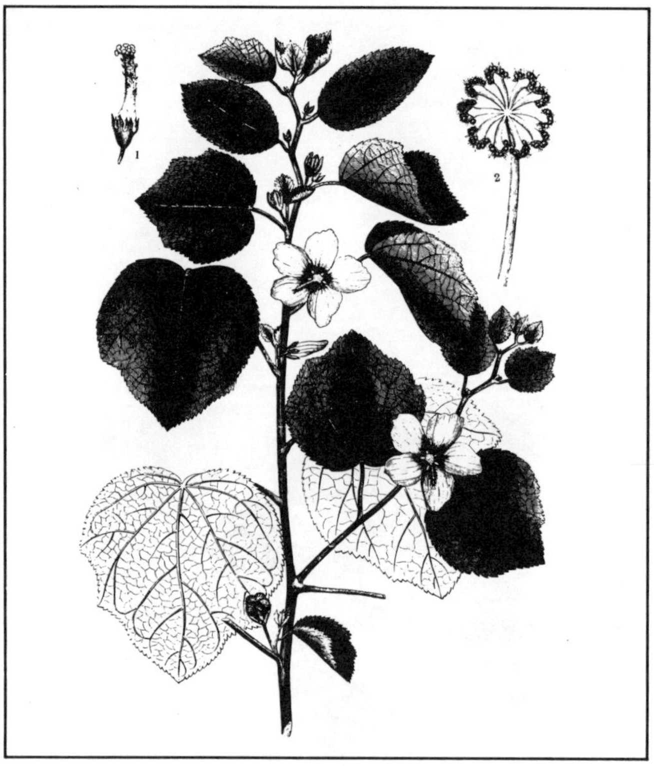

Figure 59: Urena lobata.

Urena lobata L.

Malvaceae

Common
Congo jute, Aramina fibre (Brazil), Pink burr (Australia), Hibiscus burr, Mantofu (Samoa), Qatima (Fiji), Piripiri (Tahiti).

Notes
A variable shrubby perennial. Grown as a source of fibre in West Africa and Brazil. Widespread in the South Pacific but seldom utilised. Has become a weed in Fiji, Samoa and many other South Pacific Islands. The seed capsules are covered with small spines which adhere to clothing and animal fur.

References
Ahn, P. (1963). The production of *Urena lobata* fibre in Ghana. World Crops 15, 92–97.

Purseglove, J.W. (1968). Tropical crops. Dicotyledons. Longmans Green & Co., London. pp 374–376.

Valeriana officinalis L.

Valerianaceae

Common
Valerian.

Synonyms
Valeriana alternifolia Ledeb., *V. excelsa* Poir., *V. sylvestris* Grosch.

Notes
A herbaceous perennial with a short vertical rhizome, native to Europe and northern Asia. Valerian prefers damp, fertile, slightly shaded sites, often along stream banks. Major areas of cultivation are in central Europe. The rhizome and surrounding roots are harvested in autumn and slowly dried. They are strongly aromatic and the smell affects some people adversely. Dried roots are stored in sealed containers. The major medicinal use is as a sedative (British Pharmaceutical Codex, 1963, pp 876–878). 'Valmane' is one proprietary compound containing significant quantities of valerian extract.
Other species of Valerian are utilised outside Europe including: *V. wallichii* DC. (syn. *V. jatamansii* Jones), Indian valerian grown on the borders of Kashmir and Afghanistan and used both in medicine and perfumery.
V. fauriei Briq. (syn. *V. officionalis var. latifolia* Miq.), Japanese valerian or kanoko–so, native to mountain areas of east Asia and official in the Japanese pharmacopoeia.

References
ITC (1982). Markets for selected medicinal plants and their derivatives. International Trade Centre, UNCTAD/GATT, Geneva. pp 172–177.

Wallis, T.E. (1967). Textbook of pharmacology. 5th edn. J. & A. Churchill, London. pp 360–363.

Valerianella locusta (L.) Laterr.

Valerianaceae

Common
Corn salad, Lambs' lettuce, Mache (France and Canada).

Synonyms
Valerianella olitoria (L.) Pallich.

Figure 60: Valerian.

Notes
A short (c. 10 cm) European ephemeral with tufted leaves like a small cos lettuce. A useful winter salad plant. Cultivars are rather larger than feral forms and come in two basic types, round and long leafed. Numerous cultivars are available in Europe but the species is rarely used in New Zealand. Seeds are relatively large (4 mm diam.) and germinate readily at temperatures around 5 degC. Though plants are perfectly hardy they are best grown with some protection to avoid soiling the leaves. Bolting can be prevented by maintaining temperatures above 20 degC at all times.

References
Prescott, P. (1983). A connoisseur's garden green – corn salad. Organic Gardening 30(10), 44 and 46.

Larkom, Joy (1977). Some Belgian vegetables. Garden 102(1), 21–25.

See also: Gorini, F. (1982). Ortaggi di foglia 2.53 Valerionella o dolcetta o gallinella. Informatore di Ortoflorofrutticoltura 23(10), 3–5.

Vanilla fragrans (Salisb.) Ames

Orchidaceae

Common
Vanilla, Vanira (Tahiti).

Synonyms
Vanilla planifolia Andrews, *Epidendrum vanilla* L., *Myrobroma fragrans* Salisb.

Notes
An indeterminate climbing orchid with stems reaching 5 m in height. Native to central America but is now mainly cultivated in island territories principally Madagascar and to a lesser extent in the Comoros, Reunion, Indonesia and French Polynesia. In Tahiti an additional species *V. tahitensis* J.W. Moore is also cultivated and produces pods (vanilla beans) with a quite different aroma. Though most authorities have assumed this species to be native to Tahiti, Petard (1986) suggests that it is in fact a hybrid between *V. fragrans* and the West Indian vanilla, *V. pompona* Schiede.
Synthetic 'vanillin' derived from waste sulphate liquor from paper mills and from tar extracts, has made inroads into the markets for the natural product. But there are indications of increasing demand for vanilla beans once more.

References
Purseglove, J.W. et al. (1981). Spices. Longman Group, London. Vol. 2, pp 644–735.

Davis, E.W. (1983). Experiences with growing vanilla. Acta Horticulturae 132, 23–30.

Uhe, G. (1974). Wayside plants of the South Pacific. Stocton House, Auckland. pp 22–23.

See also: Petard, P. (1986). Plantes utiles de Polynesie. Editions Haere Po No Tahiti. pp 123–127.

Vernonia cinerea (L.) Less.

Compositae

Common
Purple fleabane, Iron weed, Daudopala (Hindi), Tagulinaw (Philippines).

Synonyms
Conyza cinerea L., *Senecioides cinerea* (L.) O. Kuntze, *Vernonia albicans* DC.

Notes
An erect, slender, annual herb with lax heads of small mauve flowers, native to Southeast Asia but a widespread weed in the South Pacific and Australia. Once recorded in North Auckland. All parts of the plant including the seeds are used in Southeast Asian folk medicine. The related *V. anthelmintica* (L.) Willd. is included in the Indian Pharmaceutical Codex (1953).

References

The Wealth of India – Raw Materials. Volume 10. CSIR, New Delhi, 1976. pp 448–449.

Quisumbing, E. (1978). Medicinal plants of the Philippines. Katha Publishing, Quezon City. pp 1003–1004.

See also: Perdue, R.E. et al. (1986). *Vernonia galamensis,* potential new crop source of epoxy acid. Economic Botany 40(1), 54–68.

Vetiveria zizanoides (L.) Small

Gramineae

Common
Vetiver grass, Khas khas (Hindi), Muli muli (Fiji).

Synonyms
Vetiveria odorata Virez., *Andropogon muricatus* Retz., *A. nardus* Blanco, *A. zizanoides* Urban.

Notes
A tufted perennial grass native to India, but widely cultivated throughout the tropics. Introduced to most Pacific Islands and common in the drier areas of Fiji. An aromatic oil is distilled from the roots and used in perfumery. The foliage is woven into mats and screens but has no aroma.

References

Nair, E.V.G. et al. (1982). Review of work done on vetiver at the Lemon Grass Research Station, Odakkali. In: Atal and Kapur (eds). Cultivation and utilisation of aromatic plants. CSIR, Jammu–Tawi. pp 427–430.

Purseglove, J.W. (1972). Tropical crops – Monocotyledons. Longman, London. pp 297–298.

Vigna angularis (Willd.) Ohwi and Ohashi

Leguminosae

Common
Adzuki bean, Adsuki bean, Azuki (Japan).

Synonyms
Phaseolus angularis (Willd.) W.R. Wight. *Azukia angularis* (Willd.) Ohwi.

Notes
A small grain legume with yellow flowers. Feral forms in East Asia are scandent but cultivars are invariably 'bush' types. The seeds are angular in shape and mostly coloured dark red though other colour forms occur. The cotyledons are sufficiently firm to be ground into a flour which is used by the Japanese to make cakes. Trials with adzuki beans were conducted in New Zealand between 1969 and 1975 but the only market, Japan, proved too volatile to exploit.

References

Palmer, J. (1972). Some problems with the introduction of a new bean crop to New Zealand. Proceedings of the Agronomy Society of New Zealand 2, 65–71.

Sacks, F.M. (1977). A literature review of *Phaseolus angularis* – the adsuki bean. Economic Botany 31(1), 9–15.

Vigna mungo (L.) Hepper

Leguminosae

Common
Black gram, Urd (Hindi), Woolly pyrol (West Indies).

Synonyms
Phaseolus mungo L.

Notes
A short, (usually under 50 cm.) branched, hairy annual herb. The black or dark olive seed is perhaps the most important pulse in India where it is split (dhal) or ground into a flour. Urd was probably introduced to the Pacific Islands by Indian migrants and is established as an adventive in Fiji.

References

Duke, J.A. (1981) Handbook of legumes of world economic importance. Plenum Press, New York. pp 291–293.

Purseglove, J.W. (1968) Tropical crops – Dicotyledons. Longman, London. Vol 1, pp 301–304.

Figure 61: Mung bean plant, Otara Research Station 1975.

Vigna radiata (L.) Wilczek

Leguminosae

Common
Mung bean, Green gram, Bundo (Japan).

Synonyms
Azukia radiata (L.) Ohwi, *Phaseolus aureus* Roxb.

Notes
A small green bean best known in New Zealand as a source of 'bean sprouts'. Trials were conducted with this species at Otara in the 1970's and again at Pukekohe Research Station in the mid 1980's but yields were low partly due to bird damage. Mung bean is more extensively grown in the warmer parts of Australia.

References

Cowell, R. (ed.) (1978) Mung bean: Proceedings of the first International Symposium. AVRDC, Shanhua.

Duke, J.A. (1981). Handbook of legumes of world economic importance. Plenum Press, New York. pp 293–297.

Vigna umbellata (Thunb.) Ohwi and Ohashi

Leguminosae

Common
Rice bean, Mambi bean, Climbing mountain bean, Tsuru azuki (Japan), Anipai (Philippines).

Synonyms
Azukia umbellata (Thunb.) Ohwi, *Dolichos umbellatus* Thunb., *Phaseolus calcaratus* Roxb., *P. pubescens* Bl.

Notes
A scandent or climbing legume from Southeast Asia. Cultivated on a peasant scale from India to Japan as well as some South Pacific islands. It is both cultivated and established as an escape in Fiji. The seeds range in colour but are often dark red. In Thailand rice beans are sown with maize and grow up the stalks after the cobs have been harvested. Plants are said to be photosensitive and flower only at one time of the year (Lampang,1985).

References

Chaudhuri, A.P and Prasad, B. (1972). Flowering behaviour and yield of rice bean (*Phaseolus calcaratus* Roxb.) in relation to sowing date. Indian Journal of Agricultural Science, 42, pp 627–630.

Duke, J.A. (1981). Handbook of legumes of world economic importance. Plenum Press, New York. pp 297–298.

See also: Lampang, A.N. (1985) Agronomic aspects of legume breeding. In: G.J. Persley (ed), ICIAR Proceedings No 8, Canberra. pp 12–20.

Vigna unguiculata (L.) Walp.

Leguminosae

Common
Cowpea, Black eye pea, Southern pea (U.S.A.), Kaffir beans (South Africa).

Synonyms
Vigna sinensis (L.) Hassk., *Dolichos biflorus* Lour., *D. sinensis* L.

Figure 62: Southern pea cv. 'Purple hull', Otara, Febuary 1975.

Notes
A strong growing scandent vine. Semi determinate cultivars occur in the U.S.A. and Philippine Islands. Cowpeas are more cold tolerant than long beans and accessions from the U.S.A. grew well at Otara Research Station between 1974 and 1976. It is a characteristic of this legume that pods are borne clear above the foliage and most cultivars have a dark coloured hilum. Cowpeas are normally harvested as dry beans but in the southern United States immature beans are used both fresh and processed. This is a variable species and this note refers to ssp. *unguiculata* (L.)Verdc. as described in Verdcourt (1970) Studies on Leguminosae – Papilionoidae for the 'Flora of East Tropical Africa'. Part IV,. Kew Bulletin 24. ssp. *sesquipedalis* will be covered separately. The 'catjang' of India and Southeast Asia has been placed by Verdcourt(1970) under ssp. catjang.

References
Summerfield, R.J. et al. (1974) Cowpea (*Vigna unguiculata* (L.) Walp.). Field Crop Abstracts 27(7), 301–312.

Whittle, E.C (1984) The heat loving peas. Organic Gardening 31(1), 74–77.

Kingston, R.W. (1972). Review of grain cowpeas. Australian Specialist Conference on Crops of Potential Economic Importance, Sydney 4, 47–52.

Vigna unguiculata (L.) Walp. ssp *sesquipedalis* (L.)

Leguminosae

Common
Long bean, Asparagus bean, Snake bean, Katchang panjang (Malay), Bin (Papua New Guinea–Pidgin), Sitaw or Sitao (Philippines).

Synonyms
Dolichos sesquipedalis L., *Vigna sesquipedalis* (L.) Fruw., *Vigna sinensis ssp. sesquipedalis* (L.) Fruw.

Notes
A climbing bean with pods often 40cm or more in length. The most important green vegetable bean eaten in the pod in areas of Southeast Asia and the tropical Pacific, where the climate is too warm for the common bean. The Philippines have a particularly large number of cultivars including some semi determinate lines which require no staking. Long beans are occasionally grown in northern New Zealand but usually suffer from cold stress and exhibit yellow blotches on the leaves.

References
SAVI–Phil. (1983). A catalog of noteworthy vegetables and fruits in the Philippines. Society for the Advancement of the Vegetable Industry, Los Banos. pp 2–5.

Bautista, O.K and Data, E.S. (1976). Post harvest changes in snap cowpea under two storage conditions. Philippine Journal of Crop Science 1(4), 191–194 (cv.'Bush sitao').

Wasabia japonica (Miq.) Matsumara

Cruciferae

Common
Wasabi, Japanese horseradish.

Synonyms
Eutrema wasabi (Sieb.) Maxim.

Notes
An aquatic perennial from mountain stream sides in Japan. Cultivated for its short vertical rhizome which has a flavour similar to the European horseradish. Trials have been conducted since 1982 in New Zealand to establish an export industry to Japan.

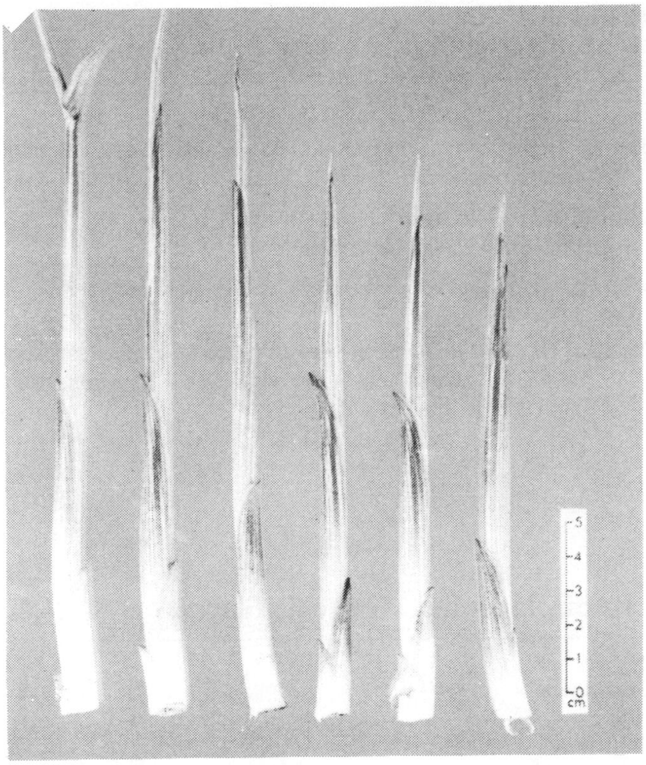

Figure 63: Myoga shoots, Christchurch, November 1986.

References

Hodge, W.H. (1974) Wasabi – native condiment plant of Japan. Economic Botany 28, 118–129.

Follett, J.M. (1986) Production of some traditional Japanese vegetables in Japan. Ministry of Agriculture and Fisheries, Hamilton. (unnumbered report), pp 1–30.

Xanthosoma sagittifolium (L.) Schott

Araceae

Common
Cocoyam, Tannia (West Indies), Yautia (Central America), Dalo ni tana (Fiji), Talo palagi (Samoa), Makue (Loyalty Is.), Tarua (Cook Is).

Synonyms
Xanthosoma edule Schott, *Xanthosoma utile* C. Koch

Notes
An aroid root crop of South American/West Indian origin, becoming increasingly common in Oceania and New Guinea as it tends to be higher yielding and has a lower water requirement than most cultivars of taro (*Colocasia* sp.) Most *X. sagittifolium* cultivars have a high oxalate content in the roots. The related *X. brasiliense* (Desf.) Engl. has been introduced to Tahiti from the West Indies as a leaf vegetable and is known by it's 'Creole' name of calalou (Barrau 1956). The same species in Hawaii is known as 'Tahitian spinach'.

References

Hackett, C. (1984). Tabular descriptions of crops grown in the tropics 4. Tannia(*Xanthosoma sagittifolium* (L.) Schott.). CSIRO Technical Memorandum 84/20, Canberra.

O'Hair, S.K. and Asokan, M.P. (1986). Edible aroids: Botany and horticulture. Horticultural Reviews 8, 43–99. Avi Publ. Co., Westport, Conn.

See also: Barrau J. (1956) *Xanthosoma brasiliense* – a useful green vegetable. SPC Quarterly Bulletin 6(1), 26.

Zingiber mioga Roscoe

Zingiberaceae

Common
Myoga (Japan).

Notes
A sterile pentaploid ginger species from East Asia. Cultivated in Japan and to a lesser extent in Hawaii. One accession introduced to New Zealand in 1980 (J.Palmer, CRD Lincoln). A further importation to Ruakura in 1986 (vide J.M. Follett). This deciduous species is quite hardy as far south as Christchurch and possibly throughout most of N.Z.

References

Palmer, J. (1984). Myoga, a possible ginger crop for New Zealand. Annual Journal of Royal N.Z. Institute of Horticulture 12, 105–108.

Follett, J.M. (1986). Production of some traditional Japanese vegetables in Japan. Restricted Report. Ruakura Soil and Plant Research Station, Hamilton. pp 31–41.

Zingiber officinale Roscoe

Zingiberaceae

Common
Ginger, Kopekai (Cook Is), Cagolaya ni vavalangi (Fiji).

Synonyms
Zingiber zingiber Forst. f.

Notes
A subtropical herb of unknown origin very widely cultivated in areas having a warm moist growing season in excess of 9 months. Most of the ginger imported into N.Z. comes from Fiji or India but there is also an extensive industry in Queensland. Many attempts have been made to grow ginger in northern New Zealand but the growing season is probably too short for economic returns.

References
Tewson, L. (1966). Australian ginger. World Crops 18(3), 62–65.

Nair, M.K. et al. (eds). Proceedings of the national seminar on ginger and turmeric. Calicut, 8–9th April 1980. Central Plantations Crops Research Institute, Kasaragod, Kerala, India. 1982, 250 pp.

Zingiber zerumbet (L.) Smith

Zingiberaceae

Common
Broad leaf ginger, Martinique ginger, Angoango (Tonga), Cagolaya (Fiji), Avapui (Samoa), Narkachar (Hindi), Awapuhi (Hawaii).

Notes
Widely cultivated in India and other parts of Southwest Asia, naturalised in Fiji and some other Pacific Islands. Used in local medicine and for flavouring. In the latter case young leaves and shoots rather than the rhizomes are utilised.
Other Zingiber species used for flavouring or medicine in Southeast Asia include:
Z. purpureum Roxb. (syn. *Z. cassumunar* Roxb.) A substitute and frequent adulterant of true ginger in India but also grown from Thailand to Indonesia.
Z. aromaticum Val. in Malaysia and Indonesia.
Z. amaricans Bl. in Indonesia.
Vide: Made Sri Prana and Sarkat Danimihardja (1981). Root and tuber crops. IBPGR Secretariat, Rome. pp 97–105.

References
Dave, R.N. (1980). Highlights of the chemistry and pharmacology of wild ginger, *Zingiber zerumbet* Smith. Agricultural Journal, Fiji 42 (1), 41–43.

Uhe, G. (1974). Wayside plants of the South Pacific. Stocton House, Auckland. p 13.

Anon (1973). Wealth of India – Raw materials Vol. XI, pp 105–106. CSIR, New Delhi.

N.B. For traditional Hawaian use of the ginger see:

Degener, O. (1945). Plants of Hawaii National Park illustrative of plants and customs of the South Seas. Edwards Bros., Ann Arbor. pp 111 – 117.

Zizania latifolia (Griseb.) Turcz.

Gramineae

Common
Asiatic wild rice, Manchurian wild rice.

Synonyms
Zizania caduciflora (Turcz.) Hand.-Mazz.

Notes
A tall growing grass widespread in wetland areas of Eastern Siberia as far south as the 'Grand Lac' of Hanoi. Cultivated particularly in South China for the swollen shoots infected with the fungal parasite *Ustilago esculenta* (vide: Zundel G.L. (1953), The Ustilaginales of the world. State College, Pennsylvania. p160). The edible portion is usually some 10 cms long by 2 to 4 cm in diameter and is marketed in leaf bases which are readily detachable. Infected clones do not flower and must be propagated vegetatively (cf. *Saccharum edule*).
The closely related North American wild rice *Z. aquatica* L. was introduced to New Zealand around the turn of the century as a source of food for waterfowl and is still to be found around the Kaipara area of North Auckland despite the efforts of noxious weeds boards wishing to destroy all vegetation not edible to sheep!

References
Stapf, O. (1909). The Manchurian water–rice. Kew Bulletin (17), 385–390.

Herklots, G.A.C. (1972). Vegetables in Southeast Asia. George Allen and Unwin, London, pp 498–499.

See also: Hance, H.F. (1872). On a Chinese culinary vegetable. Journal of Botany 10, 146–148.

For the cultivation of wild rice in the United States see: Creasy, R. (1986). The gardener's handbook of edible plants. Sierra Club Books, San Francisco. pp 291-293.

Ziziphus jujuba Mill.

Rhamnaceae

Common Chinese date, Jujube, Red dates (N.Z.), Bahir (Hindi–Fiji).

Synonyms
Rhamnus jujuba L., *Ziziphus vulgaris* Lam., *Z. sativa* Gaertn.

Notes
A spiny shrub, occasionally unarmed, grown in Northwestern China for the fruit which is usually marketed in the dried state. Jujube trees are rare but widespread in New Zealand but more common in Australia. Some Chinese trees have been reported as being up to 1000 years old. Jujube fruit is often prescribed in oriental medecine and is useful in the prevention of dental caries.
The related *Z. mauritiana* Lam., the 'ber' of India, also occurs in Australia but requires a more tropical climate. Both species have been introduced to Fiji.

References
Lyrene P.M. (1979). The jujube tree. Fruit Variety Journal 33, 100–104.

Alexander D. Mc E. (1984). Rhamnaceae. In: Page, P.E.(ed.) Tropical tree fruits for Australia. Queensland Department of Primary Industries, Brisbane. pp 161–163.

Creasy, R. (1986). The gardener's handbook of edible plants. Sierra Club Books, San Francisco. pp 144–146.

N.B. for Z. mauritiana see:
Morton, J.F. (1987). Fruits of warm climates. Creative Resource Systems Inc., Winterville. pp 272–275.

Index

— A —

— B —

— C —

— E —

— H —

— I —

— J —

— K —

— L —

— T —

— X —

— Y —

— Z —